Plant Project Engineering Guidebook

for Mechanical and Civil Engineers

Morley H. Selver, P.Eng.

Revised Edition

Multi-Media
Publications Inc.

Oshawa, Ontario

**Plan Project Engineering Guidebook for
Mechanical and Civil Engineers**
by Morley H. Selver

Managing Editor: Kevin Aguanno
Typesetting: Charles Sin
Cover Design: Troy O'Brien
eBook Conversion: Agustina Baid

Published by:
Multi-Media Publications Inc.
Box 58043, Rosslynn RPO
Oshawa, ON, Canada, L1J 8L6

http://www.mmpubs.com/

Hardcover   ISBN-13: 978-1-55489-099-6
Adobe PDF ebook   ISBN-13: 978-1-55489-100-9

Published in Canada. Printed simultaneously in the United States of America and the
United Kingdom.

Library of Congress Cataloging-in-Publication Data

Selver, Morley H.
  Plant project engineering guidebook for mechanical and civil engineers

Includes index.
ISBN 9781554890996 (bound) – ISBN 9781554891009 (electronic)

Industrial project management. 2 Engineering – Management. I. Title. II. Series.

DEDICATION

This book is dedicated to my wife Rosemary and to my sons Craig, Barry, and Jeffrey. Without their support and patience I would never have acquired the knowledge and experience necessary to write this book.

iv

ACKNOWLEDGEMENTS

The author would like to thank Mr. John Bringas, P.Eng. and the staff at CASTI Publishing Inc. for their assistance in editing and publishing the first edition of this book. The author is deeply indebted to them for their valued support, patience, and advice.

AUTHOR'S NOTE

The material presented herein has been prepared for the general information of the reader and should not be used or relied upon for specific applications without first securing competent technical advice. Nor should it be used as a replacement for current complete engineering standards. In fact, it is highly recommended that current engineering standards be reviewed in detail prior to any decision-making.

While the material in this book was compiled with great effort and is believed to be technically correct, the publisher and its staff do not represent or warrant its suitability for any general or specific use and assume no liability or responsibility of any kind in connection with the information herein.

Nothing in this book shall be construed as a defense against any alleged infringement of letters of patents, copyright, or trademark, or as defense against liability for such infringement.

PREFACE

This guidebook was written for new plant project engineers and for foreign engineers trying to understand how projects are managed in North America. New engineers joining the ranks of plant project engineering have to learn a lot of procedures, work methods, and absorb a lot of information for managing small plant projects. To the new engineer this can be a confusing learning process with a lot of trial and error. At the time of this writing, there are no books or guidebooks published to help the plant project engineer in the management of their small plant projects. There are numerous books written on project management, but these books assume that one knows the basics of project engineering, covers subjects that are not really relevant at the plant working level, do not cover material that is required, and are too theoretical for small projects.

For the thousands of foreign companies supplying services and equipment into the North American market, it is difficult and expensive to determine how plant projects are managed in North America. The companies, if able, will send engineers to North America for several years to learn the customs and management methods. However, this only trains a very small portion of the companies' workforce and there is still no reference or guidebook for the engineers back home.

I started working as a plant project engineer in a pulp mill then moved on to central engineering for a large industrial corporation. There we managed large projects in the 11 plants that we owned. I noticed that plant project engineers were not very well educated in the field of plant project engineering as there were no proper training programs available. It was this lack of training that made the company have a central engineering department to manage their projects.

As I moved around and my career progressed in the area of plant project engineering I had the opportunity to develop and use contract documents, set up plant engineering departments, and manage plants. All this gave me a good understanding of what basic knowledge a plant project engineer should have. I have written this book to give the new plant project engineer background knowledge of the project authorization process including budgets and estimating, information on how to control the office and drawing functions, a description of the bidding and procurement process, the basics of construction management, contracts, commissioning procedures, startup, and training. This book is not theory, as the information and forms can be taken from the book and actually used in your projects. Although this book is written in Canada, the information has been used in both Canada and the US. The information is common to the oil and gas, pulp and paper, board and recycling industries.

It is my hope that plant project engineers and foreign engineers will find this guidebook useful in understanding the issues and methods necessary to successfully manage their plant projects.

Morley H. Selver, P.Eng.

TABLE OF CONTENTS

Chapter 9

Chapter 10

Chapter 13

I keep six honest serving men (they taught me all I knew);
their names are **WHAT** and **WHY** and **WHEN**
and **HOW** and **WHERE** and **WHO.**

Rudyard Kipling

Chapter

1

INTRODUCTION

Why This Book

Industrial plants are the training grounds for young project engineers, who are usually hired by the plants to fill an immediate need for help. On the first day, a coworker will give them a tour of the plant and office. They will be introduced to a lot of people they will need to know whose names will be forgotten in minutes and will be shown to their new desk. There they will find all the information required to perform their job. It is expected that they will "learn on the job" and they generally do. There is nothing wrong with this approach and it usually works quite well; however, there is a difference between learning and understanding. This book is intended to steer the project engineer toward an *understanding* of what they are trying to "learn on the job."

This book is written from the perspective of a plant project engineer and is based on a typical plant engineering, procurement, and construction contract (EPC). However, the principles and methods discussed also apply to other engineering areas in all industries. All projects go through the same steps and require the same information to be implemented. For small projects handled in-house, this book will give the project engineer an understanding of why certain things are done, and for large projects where outside consultants are involved, it will give an understanding of what the consultants do and what the project engineer should be looking for. Most courses and books about project management deal with large projects and assume that the basics of project management and project control are understood. This is not necessarily the case. The majority of everyday projects encountered in a plant are fairly small (less than $5 million), and using project management techniques designed for large projects can be cumbersome, time-consuming, and not necessarily cost effective. By understanding what is required for each project and why, a reasoned approach can be taken to employ the right techniques to get the job done without being burdened with questionable tasks developed specifically for large projects.

This book covers the basics of what a project engineer should look for when gathering information for the Project Approval Phase, pitfalls to avoid in the Design Phase, what to ask for in Request for Quotations, and problems which will be encountered during the Construction Management Phase. By using this book as a guide, the Project Engineer will get enough information to control and manage projects effectively. The aim is for the project engineer to have a clear understanding throughout the project life cycle of **WHAT** is to be done, **WHY** it is being done, **WHEN** it will be done, **HOW** it will be done, **WHERE** it will be done, and **WHO** will do it.

The information in this book comes from knowledge gathered during my 20 plus years of working for operating pulp and paper mills, in corporate central engineering, and as a consultant in the pulp, paper, and board industry. This book is intended to help project engineers perform their job in an efficient, professional manner by understanding and maintaining control throughout all phases of their projects.

General Guidelines

The following are some general guidelines to keep in mind when managing projects:

1. It is important to keep your supervisor informed to the fullest extent throughout the project. Just like you, your supervisor does not like surprises.

2. Treat all projects as if there will be a legal court case. This action may help avoid one. Take meticulous notes, make sure that all information in and out of your office is correct, and keep records of all in/out transactions.

3. Purchase a scribbler (notebook) and keep all handwritten meeting minutes pertaining to a project in it. This way all the minutes are in one place. This takes care of the problem of loose sheets of handwritten minutes floating around and getting lost. Somewhere, sometime you will have to refer to your handwritten notes.

4. Always have someone else take minutes at meetings so that you, as the chairperson, can concentrate on the subject at hand and not miss something important while trying to write minutes.

5. When managing projects, there will be stages when the workload gets excessive and stressful. The usual tendency is to rush the work and do as much as possible in the shortest time. At these times do not panic. Pace yourself to be fast but steady and avoid shortcuts. This will result in a better job and will be less stressful.

6. Do not depend on memory. Use a scribbler, or a company supplied journal, keep very good notes and a daily diary. As a minimum, everything that transpires during the day should be recorded.

7. Take a course in technical writing or report writing so you can write all correspondence in clear, concise language that is to the point with as few words as possible.

8. Each industry has its own abbreviations that are used daily without explanation. To avoid confusion when writing, spell out the words in the abbreviation at least once, typically the first time it is used.

9. Set up a tickler file. This is a filing system you use to follow up on assignments, i.e., when you ask someone to do something by a certain date, you want to be able to follow up on that date. One easy Tickler File[1] I currently use is to make 31 file folders and label them "1" to "31"—one for each day of the month. Label 12 hanging folders each with a different month. Put the hanging folders in your filing cabinet drawer and insert the 31 file folders in the hanging folder for the current month. As each day passes and records are collected for that day, take that day's folder and move it to the next month's hanging folder. Work your way through the month moving the day file folders to the next month's hanging folder. Make a note of items due, or use a photocopy, and place it in the appropriate day folder. Every morning when you come in, look in the current day folder and you will see what is due that day. This tickler file surpasses the computer and personal digital assistant (PDA) methods of tracking as you only have to look in one place for the information and not two (the computer and the file).

10. Be honest, fair, and ethical in all your dealings. You want to be able to sleep at night. Don't get yourself in compromising situations that may force you to cover up or act defensively in the future. Use common sense. If someone gives you a gift that might in anyway appear to influence your decision making on the job (for example, a vendor who sends you a valuable Christmas gift), consider donating the gift to a local charity. Get a receipt for the donation and note it in your records. You need a clear conscience to act effectively to protect your employer's interests.

---

[1] Taylor, Harold L., "Personal Organization: The Key to Managing Your Time and Your Life," Time Management Consultants, North York, ON, 1998.

Chapter

2

PROJECT AUTHORIZATION

Plant Capital and Maintenance Budgets

Plants, in fact all companies, have long range plans and budgets. In industrial plants there are usually three budgets—a capital budget, a maintenance budget, and an overhead budget. The capital budget covers new equipment or improvements to the plant. The maintenance budget takes in repairs to existing plant equipment or structures. The overhead budget covers services provided to the plant such as accounting, stores, etc. In most jurisdictions, items on the capital budget are taxable, while maintenance budget items are a tax write-off against the plant operations. The maintenance budget costs and overhead budget costs are part of the plant operating costs. Therefore, the maintenance budget can have significant effect on the plant's profit and/or loss.

The budgets contain line items regarding new equipment or structures (capital budget), maintenance work required on equipment or plant structures (maintenance budget), the projected cost to supply service to the plant for the upcoming year (overhead budget).

The three budgets are developed and submitted to management for approval, usually once a year; however, they are worked on throughout the year. The budget line items will have a dollar value assigned that is a rough order of magnitude (ROM). The dollar amounts assigned to the line items could come from department managers, supervisors, engineers, or are sometimes amounts left over from previous budgets. Some ROM estimates are pure guesswork and others are determined by talking with vendors about equipment prices and getting estimates of installation costs. The time spent on the each estimate will vary and each estimate's accuracy will vary but should reflect actual anticipated costs as much as possible.

For a ±10% estimate, the backup information you use has to be detailed, current, and accurate. This requires a lot of up-front engineering work. In fact all the detailed engineering should be complete and current pricing obtained using detailed engineering drawings. The pricing would include material suppliers, vendors, and contractors. This type of estimate requires the company to spend money up-front to dedicate a person to the project, do the detailed design, obtain and evaluate proposals, and prepare the estimate. Although this is the preferred type of estimate, there are typically very few of these estimates.

At the midrange of ±20% to ±30% estimate, you would have a preliminary flowsheet and layout and a basic scope. You could be using past equipment pricing with a multiplier or a preliminary price from a vendor or contractor (with his own accuracy range). You have looked at the installation requirements and prepared a lump sum estimate based on them. You may get installation information from published directories. The less information you have, the more your estimate will lean toward ±30%. Most budget items will be in this range.

At the end of the scale is ±50% estimate which is based on very little information and a lot of experience. This would take very little time to prepare. The less experience you have the more the estimate will be off and a range value higher than ±50% may have to be used. A lot of the budget items would be in this category.

Assuming the plant is one of many in a corporation, plant management will scrutinize, modify, remove, and prioritize budget items to get the budget within guidelines provided by the head office. One of these guidelines for capital budget items is *Rate of Return*. The benefits of installing the equipment or structure are looked at and a dollar value assigned to the benefits. Examples of benefits would be:

- Cost savings due to less manpower required to perform the function;
- Improved quality would allow selling product for a higher price;
- Speed improvement would allow more product to be sold;
- Savings in maintenance costs;
- Savings in cost of raw materials;
- Ease of use decreasing operator errors.

Each line item will have a benefits value associated with it and these will be evaluated further at the project approval phase.

The Rate of Return is calculated as follows:

$$((Benefits - Cost) / Cost) \times 100$$

A Rate of Return of 10% is typical; however, a value as high as 20% has been used. Every plant is different and the current state of the plant business will determine the required Rate of Return. For some projects the benefits are fairly easy to determine and cost. For other projects the benefits are nebulous and have to be searched for. For these borderline projects there will be pressure to either lower the estimated project cost or to increase the benefits. Be very careful with your estimates when dealing with these types of budget items.

Once the plant is satisfied with their budget, they will then submit the total budget dollar estimates to the head office for approval. At the head office the budget is usually pared back, discussions held, and the budget reworked until an agreeable dollar value is reached and the budget line items are approved. Line item and budget approval usually depends on how the market is for the plant's product:

- If economic times and profits are good, then a lot of capital projects as well as maintenance items will get approved.

- If times are bad and profits mediocre, then less capital and more maintenance projects will get approved.

- If times are really bad with no profits, then no capital projects and only some maintenance projects get approved.

- If times are extremely bad and profits are not on the immediate horizon, then only essential maintenance items will get approved or no items at all.

If the plant is the only one in the company or if there are only a few plants in the company, there is not as big a demand for the pool of budget money available and the chances of getting the budget line items approved are reasonably good. If the company owns many plants, there will be only one overall Capital and Maintenance Budget for the whole company. All plant budgets get put into the pool and the head office will then determine what the total budget will be and where it will be spent.

Sometimes a large expenditure has to be made and, if known in advance, the expenditure can be planned over a few years. This will usually happen when a major piece of equipment has to be replaced or process improvements made. The Rate of Return is looked at and plans made for the expenditure. These projects will sometimes take up a large portion of the plant's budget and leave little money for other line items. If the plant has a major catastrophe and is required to allocate a large amount of money to maintain production, then the next year's budget may be cut and/or important work put off. Rather than exceed the overall budgets, approved funds will, if necessary, be shifted from one plant to another to accommodate major problems. These are decisions management has to make in the best interest of the company.

One other consideration for budget item approval is the issue of cyclical markets. An example is the pulp and paper industry, which operates on a seven year cycle, i.e., over a seven year period a plant could expect to go from unprofitability to profitability to unprofitability. If the market for your product is down (low part of the cycle) and capital budgets are lean, can the money be spent now on a major process improvement so the start-up of the process coincides with a market upswing? The reasoning behind this consideration is that possible savings in equipment and labor costs can be realized during the downturn. Equipment savings of 50% are not unheard of, and when you are dealing with equipment worth several million dollars, the savings can be substantial. If you are doing a large project with lots of manpower, the people are available during downturns and you do not have to pay the premiums necessary during boom times. These market swings are hard to predict today, although some companies do attempt to predict them. In the 70s and early 80s it was possible to predict two years in advance what part of the cycle the market would be in. Projects were actually planned based on a plant being shut down during a given month two years hence because of a predicted downturn in the market.

Case History 1

I worked in Central Engineering for a large paper company that operated 11 different plants across the country. Their core business was newsprint; however, they had acquired plants and business in other products and industries. During one year the total Capital and Maintenance Budget for all plants was $250 million. If divided evenly that would be ~$23 million for each plant. However, some plants needed repairs more than others did so they got more than the $23 million. Some plants were in specialty markets that were booming, so they got more of the moneys to increase production. The rest of the plants got a lot less of the pool of money or nothing at all. By having to dilute the budget money available for their core business, the viability of some of their core plants was affected. The only way to protect their core business was to divest themselves of their no-core businesses. This dilution of capital and maintenance money is one reason you find companies divesting of their newly acquired operating divisions that were never part of and have nothing to do with their core business.

Once the budget line items are approved, a contingency is added and the total budget is returned to the plant for project execution. The contingency can sometimes be substantial since it is used to cover costs associated with unplanned events. The line item numbers can become project numbers or the plant may have a different numbering system. The accounting group issues and controls the project numbers.

Although the budget line items have been approved that only means that a pool of money has been allocated for those items. The budget line items have to be looked at in greater detail and submitted for formal approval to plant management, or depending on the project cost, to the corporate head office. All plants have management signing levels. The department manager may be able to authorize $25,000.00 for example. If the project cost is greater than this, the next level of management has to authorize the project. An upper limit for most plant managers would be $100,000.00. Projects costing more than this would require corporate head office authorization. During the formal approval process the line item could be rejected if:

- Detailed study results in a cost that is higher than budgeted;
- The project proves to be unfeasible;
- There is an unplanned expenditure in the plant and its priority is higher than the line item's priority.

The following section covers investigating a budget line item and submitting it for management approval.

Project Authorization and Scope Documentation

Your supervisor has assigned you a budget line item to work on and submit for management approval. Attached to this budget line item request should be a charge number relating to the budget line item. As described in the previous section, this budget line item is a preliminary estimate only. Your job now is to look at the budget line item in greater detail, complete the proper documentation (listed below), and submit it for management approval.

The project approval documentation usually involves four basic documents:

- a scope of work
- a cost estimate
- an overall project schedule
- a spending forecast.

Most companies will have various forms for these items and you should use them. As the project engineer you would complete the above documents while the accounting group would prepare other items such as the Rate of Return and Financial Analysis with some input from you. All the forms are fairly standard and are used throughout North America, across all industries. Each company may ask for the odd item that is specific to that company, but in general company to company, industry to industry, the same information is required. The form shown in Figure 2.1 is a typical Project Authorization form containing the items that should be covered for the approval phase (detailed discussion follows).

Project Authorization

Date:	PA #:	
Project Title:		
Prepared By:		
Department:		
Area Specialist:		
EWR Number:		
Work Order Number(s):		
1.1 Project Objective:		
1.2 Schedule:		
1.3 Critical Assumptions:		
1.4 Drawings:		
1.5 Standards, Specifications & Codes:		
1.6 Process Requirements:		
1.7 Battery Limits:		
1.8 Mechanical/Civil/Structural Requirements:		
1.9 Electrical Requirements:		
1.10 Instrumentation Requirements:		
1.11 Safety:		
1.12 Training, Startup, & Commissioning:		
1.13 Permits:		
1.14 Environmental:		

Figure 2.1 Project Authorization Form – Scope

The time period for preparing the documentation can vary from 24 hours to six months or longer depending on the size and the urgency of the job. Common sense dictates that when preparing documents and estimates, the more time spent up front on the study, the more accurate the estimate and the more defined the scope will be. It pays to get proper estimates from contractors for small jobs, and for larger projects it may be prudent to hire a consulting engineer to do some up-front engineering.

As much information as possible should be included in the written documents. Even the smallest items should not be left out—write them down somewhere. If the documentation only allows for a summary, then details should be written out and filed for later retrieval. This is the only way you will be able to determine exactly what was included once approval is obtained.

Sometime in your career you will be asked to prepare a rush estimate for a project and you will be told that approval will be "no problem." Your tendency will be to rush the information gathering process, taking shortcuts and thinking that all will be remembered when approval comes because approval is just a formality. Often these are projects that:

- do not get approved right away;
- come back to you with a request for something slightly different or with a demand for more for less.;
- raise questions you do not have the answers for but should have.

In other words—**TROUBLE.** That is why, when preparing approval documents, you need to put everything in writing, make good notes, and do not count on your memory. With all the other items you have to work on, how much are you going to remember and understand about the concept six months from now? Even during the approval process you may be asked for a clarification, so you have to be prepared with the correct answer. No matter how much written information you have, there will be times when what has to be done is still not clear. If you do not have anything in writing, it is even more difficult to clarify items. This may seem trivial but starting a project and not knowing what is included will put you behind the eight ball at the beginning, which is not where you want to be.

Also, someone else may be given your project and you would have to explain to him or her what the scope is. Would you be able to do this? How would you like to take over a poorly defined project? Read the information you have noted and ask yourself the following question: In one month will I be able to describe, in detail, to someone exactly what is included and required? How about in six months? We all know that information is power, but in this case the information you have assembled should be readily available and understandable by others. Do not set up a system only you can figure out. This is a bad business practice.

A useful tool for organizing your project estimates is the Work Breakdown Structure (WBS). The WBS has been around for a long time and for good reason. It forces you to list all the items that have to be done, assigns a cost to each item, and is your backup as to what is or is not included. Consider the purchase and installation of a pump as outlined in the following WBS example shown in Figure 2.2 and 2.3.

If possible, your estimate should be for a clearly defined item; i.e., a specific model with all items that are included with it. If a question arises as to an item's inclusion, it is a simple matter of looking it up in your estimate sheets.

Keep in mind that you are the leader of the project and as such you have the overall picture or world view of what is actually going on and what is involved. The majority of others in the system will have no understanding as to the driving forces for the project. They may not have the same education as you, and in order to perform their function, they will have to have information at a very basic level. This requires information to be spelled out clearly, precisely, and in simplified terms using tried and true methods. One of these is having your estimates organized via the WBS mentioned above. Do not take shortcuts combining work items, as it will only increase your workload when you have to sort it out at a later date. Keep it as simple as possible.

Item	Quantity	Unit	Description	Material	Labor	Total
200			Supply & Install Pump			
200.01			Purchase Pump			
.1	2	Hrs	Spec Pump		2 Hrs @ $50.00 = $100.00	
.2	2	Hrs	Spec Motors		2 Hrs @ $50.00 = $100.00	
.3	4	Hrs	Prepare and Issue RFQ's		4 Hrs @ $25.00 = $100.00	
.4	2	Hrs	Bid Evaluations		2 Hrs @ $50.00 = $100.00	
.5	1	Only	Purchase Pump 200 GPM, Centrifugal, 304SS	$5000.00		
.6	1	Only	Purchase Motor 30HP, 1800 RPM, 3/60/575	$1500.00		
.7	1	Hr	Receive Pump		1 Hr @ $25.00 = $25.00	
.8	1	Hr	Receive Motor		1 Hr @ $25.00 = $25.00	
			Store and Handle		In Above	
	1	Lot	Freight	$500.00		
			Contingency	$500.00		
			Subtotals For 200.01	**$7500.00**	**$450.00**	**$7950.00**

Figure 2.2 Typical Work Breakdown Structure Form

Item	Quantity	Unit	Description	Material	Labor	Total
200.02			Install Pump			
.1	1	Only	Pump Foundation	$2500.00	In Material	
.2	1	Hr	Move Pump & Motor to Site		2 M@1 Hr@$50.00 = $100.00	
.3	2	Hr	Set Pump		2 M@2 Hr@$50.00 = $200.00	
.4	4	Hr	Align Pump to Vessel & Piping		2 M@4 Hr@$50.00 = $200.00	
.5	2	Hr	Set Motor		2 M@2 Hr@$50.00 = $200.00	
.6	2	Hr	Align Motor		2 M@2 Hr@$50.00 = $200.00	
.7	8	Hr	Hook up Power	$1000.00	2 M@8 Hr@$50.00 = $800.00	
.8	1 lot		Grout Pump	$700.00	In Material	
.9	2	Hr	Check Rotation		3 M@2 Hr@$50.00 = $300.00	
.10	2	Hr	Connect Motor to Pump		2 M@2 Hr@$50.00 = $200.00	
.11	2	Hr	Commission & Start-up		2 M@2 Hr@$35.00 = $140.00	
.12	1	Lot	Tools & Equipment	$1500.00		
.13			Contingency	$500.00		
			Subtotals For 200.02	**$6200.00**	**2340.00**	**$8540.00**
			Total			**$16490.00**
					Say	**$17000.00**

Figure 2.3 Typical Work Breakdown Structure Form

Even with the best advanced planning, however, changes will be needed. Items will be left out, someone will have a better way of doing the project, or future project owners will want something more expensive. If through proper documentation and estimating sufficient funds have been approved, there may be some spending leeway. Sufficient funding always makes the project easier to manage and run. There is nothing worse than an underfunded project.

By knowing exactly what is included in your estimate it is easier to determine the requirement for scope changes when someone wants something extra added to the project. At least you can say with honesty that it is an "extra" and you'll have the backup to prove it. Management should approve any scope change requests before the work takes place. Don't get caught up in trying to keep your fellow employees happy by adding unauthorized scope changes. The scope change may not seem like much at the start of the project but it could have a major effect on the final project cost and you will have no excuse for the scope addition. Letting management say yes or no to a scope change request takes the pressure off you of going against your fellow employees' wishes.

A Project Authorization Form (Figure 2.1) is a brief summary of what is included in the project and lists the items that should be approved by management. The required items will vary from plant to plant, however, this form is a good basis to start from. The intent is to make the authorization form as complete as possible so management has all the information required to make an approval decision. Do not try to hide things management may not want to hear, as these will come back to haunt you once the project is approved. If management wants to take items out of the scope, it is their prerogative, but it is important they have all the information and backup required to make a decision. The Authorization Checklist (Figure 2.4) covers all items that should be looked at when making up the Project Authorization Form. As items are considered and completed the item can be checked off so that you know you have looked at it. You can use a Y (yes) or N (no) as applicable to have a list of what information or documentation is needed for project authorization.

This authorization documentation and accompanying estimate will be the basis for your project. Any changes to the project after approval will have to have a scope change approved by management. Therefore, it is vital that what is being approved, in the project authorization, is known in as much detail as time and available information allows. It is not only your fellow workers who may want changes, but management may also want to change something. A Scope Change document should apply to management change requests as well. You may have to watch your step so you do not upset your supervisor when you insist on acquiring the documentation; however, you will have the backup information outlining what was included.

Project Authorization Document

The following discussion explains each part of the Project Authorization form shown in Figure 2.1.

Date

This is the date you submit the form to management. Do not use any other date as this will show you how long the approval process takes. It will also give an indication of how long it took to prepare the approval document. This can be done by comparing this form date with the date on the engineering request. Both of these dates will be of interest to you for future reference so keep track of them.

PA Number

This is the Project Authorization number assigned by the accounting group and tying the project back to the budget line item. The number will be the same throughout the duration of the project. This number should be shown on all documents relating to the project as your accounting system, filing system, drawings, purchasing documents, etc. will use this number to collect information and define what project the document is for.

Project Title

The title you give the project should describe the project and have meaning to others in the plant. Titles are sometimes given that bear no resemblance to what the project is actually about. Poor project titles can be confusing and do nothing for the communication flow in the plant. For example, the project "Washer Vent Disengagement Section" would be less confusing if it was titled by its more common name of "Washer Vent Demister." You can change the project title from what is shown in the budget line item. Just make sure that you reference the old title in your authorization document.

Prepared By

This is the name of the person who prepared the documentation, presumably you.

Department

This is the department in which the project will be carried out. If the project crosses several departments, use the department that handles the major portion of the project or the department that originated the request.

Area Specialist

In large plants there are usually process engineers and area engineers who are classed as specialists in their respective areas. They will usually have a detailed understanding of the process, the equipment, and what the problem is. This specialist should work with you to prepare the documentation. He will be able to describe the project objective in great detail and assist in the scope definition. This person may move on and may not necessarily be the specialist working with you on the project. Sometimes, years later, someone needs to know who worked on the project; therefore, this information should be added. If your plant does not have specialists and you filled the roll, then put your name for this item.

EWR Number

This is the Engineering Work Request number. To control what projects plant engineers are working on, engineering departments use a Work Request System. This Work Request System ensures that the plant engineers only work on approved work. All requests for engineering assistance are sent to management for approval. If the request is approved, then it is assigned the next sequential number from the Engineering Work Request list. This keeps plant engineers from working on frivolous items that do not reflect the objectives of the plant. Depending on your plant's accounting system, your time will be charged to this number. The master list will cross-reference this EWR number back to the PA number and eventually to the budget line item. The costs collected under this EWR number will be charged back to the project upon approval.

Work Order Numbers

People outside the engineering department usually work under a work order numbering system. For a person to work on a piece of equipment or to help you with your assignment requires the issuing of a work order number to track their time. This system allows the plant to track the personnel portion of its maintenance costs. The numbers you list are the ones used to collect the costs associated with the approval documentation. These costs are usually charged back to the project once it is approved.

Because these work order numbers are like blank cheques, it is very important to keep track of them and who is charging their time to them. Since some people are always looking for active work order numbers to charge to, make sure all your work orders are closed once the task is complete. In fact, issue and close work orders often so the work order numbers do not get out into the plant and unauthorized people start charging to the number. The longer the number is out there, the more likely there will be abuse. Keep in mind cost reports are usually issued about once a month and by the time you find out about unauthorized charges it may be too late. If you find unauthorized charges, go back to the accounting group and have them removed. Because of the possibility of unauthorized charges, do not use blanket work order numbers. A blanket work order number is a number issued to collect the costs for an assortment of tasks over a long period of time.

1.1   Project Objective

This is the justification for doing the project. It should include a description of the project's purpose and what is expected to be achieved. It should outline what the existing situation is, what is to be accomplished by carrying out the project, and any special requirements or limitations that management should know about. If possible the project objective should be expressed in measurable, fixed, and identifiable terms, e.g., the project will increase production by 25 tons; the vendor's process guarantee; etc.

For example: This project will replace the non-reversing gear boxes on the Archimedes Screws at the effluent treatment plant. There are 3 screws with 2 running all the time on a 16-hour rotation. When the screw stops, the water does not drain back down to the pit, but stays in the screw. During the winter, this water freezes and puts the screw out of service until the maintenance crew can thaw the ice. These freeze-ups have occurred at least 4 times every year for the past two years at a cost of $50,000 per occurrence.

By replacing the gear boxes with a reversing type, the screws will be allowed to rotate backwards, draining the water and preventing freeze-ups. This will allow the operating rotation of units to be maintained more efficiently and save $200,000 per year in maintenance costs.

The plant's project sponsors or area specialist will usually help you with this section. It is important that this section be written in clear, concise terms. If, as the project progresses, there are attempts to add items to the project or change the project concept, this section can be referred to to see if the request is relevant and within the objective of the project. Because these issues will have to be dealt with, it is important to have this section clearly written.

1.2   Schedule

This should at least outline proposed milestone dates for the project. It should include: project approval time (this will be a guess), permit approval times, start of engineering, when major equipment has to be purchased by, delivery dates of major equipment, start of construction, shutdown dates and durations, startup date, and full production date. Larger projects will require more schedule information but the above would be sufficient for smaller projects. These do not have to be actual dates but can be indicated as number of days/weeks/months after project approval. That way it doesn't matter that the project receives management approval within a specified timeframe. However, if a fixed date (e.g., a shutdown) is driving the schedule, then the schedule should be within a specified timeframe with a "drop-dead" date for approval. Management approval is probably the hardest date to estimate. Do not be fooled into thinking approval will be immediate, no matter how important the project supposedly is. Protect yourself with a drop-dead date.

Preparing a schedule makes you think about how the project will go together and will help identify missing items. It also gets you talking to suppliers. This will give you a better feel for the viability of the project and if there are potential equipment delivery problems. You may even find there is another solution to the problem that could save money.

If you have a scheduling program, use it. It will make your document look more professional and present the information in a more readable format. But be careful with scheduling programs. Some are designed primarily for large projects and do not lend themselves well to small projects. Others are easy to use for small projects and produce Gantt charts (bar charts) quite readily. For approval schedules, milestone dates may be faster and just as adequate. The dates can be shown in Gantt chart form or simply listed depending on what format is desired by management.

Do not put too much information in the management schedule. This can be confusing and may end up in nitpicking questions and more work for you. Stick to the relevant points. If management wants more information, they will ask for it.

The only way to keep proficient at computer programs is to use them all the time; therefore, it may be worthwhile to use the scheduling programs for all your projects no matter how small. Most programs will allow you to take a detailed schedule and roll it up into a milestone schedule if that is what you require.

Scheduling should not be taken lightly. It should be relevant and not put together haphazardly. Management will always remember any favorable dates you give them even if you tell them the dates are just a guess. For this reason, it is important to specify only those dates you are sure you can meet and to make clear any qualifications affecting those dates. Most projects will tie in with plant shutdowns and these dates should be known well in advance. On larger projects, plant shutdowns scheduled a year in advance may determine your schedule or, conversely, your project may determine the shutdown date and length. As you get more experience in managing projects, you will find that the end date never changes even when the front-end dates cannot be met because of late management approval or other problems. Your schedules will always be compressed and you will spend a lot of time protecting the dates. For every change to the project, you should look at the schedule impact, and if an extension is required it should be shown on the schedule.

1.3   Critical Assumptions

All assumptions used in developing the budget estimate should be listed. Because your assumptions affect the final estimated cost, it is important that you and your management understand how the estimate was derived. Someone with more experience may advise you that some of your assumptions should not be used or may add other assumptions. Management is usually looking at the big picture and does not look at the details. The assumptions will get management thinking about different aspects of the project and how it could affect other parts of the plant.

Once the project is approved, all your assumptions will be part of the project basis. Throughout the project, if changes occur such that any of the assumptions no longer apply, then the project budget estimates and schedule deadlines should be changed accordingly, e.g., additional funds should be allocated to compensate for the monetary change and the schedule should be extended to compensate for the time required to do the extra work. Management may and often does refuse the additional compensation for small items. For bigger items there may not be any additional compensation until the end of the project when all the costs are in. If there is a budget underrun, the funds would go to pay for the changes. If not, the additional funds would have to come out of the pool of project money. Keep records of what transpires to help explain any overruns at the end of the project. Assumptions can be as simple as:

- contractor or plant forces to do the work
- union or non-union contractors
- availability of special equipment or material
- suitability of existing foundations
- availability of permits
- sufficient quantity of utilities
- reuse of existing plant equipment

1.4   Drawings

List the drawings and sketches used to develop the estimate. The drawings could be original plant drawings. These drawings, in the end, may be obsolete but they at least show what the project approval was based on. All drawings should be numbered, dated, and uniquely identifiable in some way. There should be as a minimum a flowsheet (if applicable) showing how the project will tie in with the existing process. Submit this flowsheet with the approval package.

By going through the exercise of producing drawings or sketches, other things pertaining to the project will come to light. This could lead to further assumptions or additional items that have to be covered in the project. When doing field research for information to produce a drawing or to confirm information, the area in question may not be accessible because the plant is operating. In this case certain assumptions about the object will have to be made.

Case History 2

We had to install a diverter gate in a plant process sewer. When the plant was up and running, the presence of H_2S gas made the area inaccessible. In order to estimate the project cost we had to assume the drawings in the file were correct. We protected ourselves by ordering the equipment in such a way that any errors in the trenches could be made up with grout between the equipment and the trench.

1.5   Specifications, Standards, and Codes

As with drawings, list any and all specifications and standards used in making the estimate. These could be in-house specifications, consultants' specifications, government specifications, industry standards, or vendor specifications. Because the approval documentation does not go outside the company, most any specification that meets the project needs can be used. (This means you can use another company's specification for the package.) If the project is approved, the specification can be revised or rewritten on the plant letterhead to reflect the current need. The specifications should be numbered, dated, and uniquely identifiable. Again, this process can bring to light project items that may need further clarification or assumptions to be made.

Upon project approval you will have a ready list of specifications and standards to be used. Sometimes you will get approval and have to start work immediately. You could be hard pressed to find the time to search for appropriate standards and specifications. This can save you a lot of valuable time.

Codes that apply to the project should be listed in the appropriate section on the authorization form. This could include building codes, piping codes, tank guidelines, National Fire Protection Association (NFPA) codes, etc. Be aware that there may be local requirements for any of the codes and you should talk to the local inspector (this could be the building inspector, boiler inspector, plant authorized inspector, or area fire marshal to name a few). Some of these codes can affect the schedule due to the length of the approval process. Each jurisdiction has its own rules and regulations that have to be followed.

Case History 3

We had purchased some used pressure vessels and moved them from one jurisdiction to another. Before we could install them they had to be retested and approved by the local Boiler Inspector. The high-pressure vessels passed, but the low-pressure vessels failed. We had to test one of the low-pressure vessels to destruction in order to pass all of them. It failed at less than the required test pressure. We then had to come up with alternative pressure vessels. This was something we took a chance on and lost.

1.6   Process Requirements

Describe what product is to be made in terms of quality and quantity, being sure to include any specifications, tolerances, and process guarantees.

For example:

- you could be improving the system so the final product meets a published standard, or
- the new equipment will allow your product to be produced with a surface tolerance within ± 0.002", or
- the equipment supplier guarantees that X tons of raw material into the process will result in Y tons of finished product out of the process.

You should also include the quantity requirements needed to make the product. Some of these would be:

- the amount of chemicals used,
- how much additional raw material is required
- the additional utilities required such as power, water, compressed air.

If the project results in using less of any of the input quantities, these savings should be noted as well. A flowsheet or a mass flow block diagram showing how the new process will mesh with the existing system should be produced and any flowsheet or mass flow block diagram should be balanced, i.e., the material flow out of a piece of equipment should equal the material flow into the piece of equipment.

Often someone will try to expand on the savings generated by using less of a certain utilities. These savings are really hard numbers to confirm and to project into the future. Management will be very skeptical of any large savings, so be prepared with believable backup to prove your point.

1.7   Battery Limits

The battery limits are the boundaries of the project. They should be well defined, especially in an operating plant. This is done by identifying all the starting and ending points of the process and services associated with the project. These boundary conditions apply to all services (electrical, sewage, etc.), structural work, and civil work, if they are part of the project. Examples would be:

- The line will start at Valve 123 on Tank ABC and run to Nozzle 111 on Tower XYZ.
- The area of work is bound by the column lines E12, N17, S15, & W 10.
- Excavation will extend from a point 25 feet east of the road centerline to a point 100 feet east of the road centerline.

This section should describe clearly what the battery limits of the project are. You will have to know what these battery limits are when you start dealing with contractors or plant forces. If the limits are too difficult to describe, a sketch can be made showing them.

1.8   Mechanical/Civil/Structural Requirements

This should be a summary of mechanical/civil/structural work or equipment included in the project. Examples would be:

- move 30 yards of dirt
- install 4 tons of miscellaneous steel
- install seven pumps and associated piping
- relocate operator's booth

Project size will determine the detail required.

1.9   Electrical Requirements

Summarize what electrical work and/or equipment is included in the project. Electrical power usage can be a detriment to the approval of projects and can be the reason some projects are never done. Some plants buy power in blocks and additional horsepower can put the plant in a more expensive block, thereby affecting the plant operating costs. Some questions that should be addressed in this summary are:

- Will this project affect power costs?
- How will the project affect the plant power requirements?
- What affect will the project have on existing plant electrical equipment?
- Does the project present any risks to the plant power grid?
- Are outside power and government agencies involved and if so how?
- Are electrical shutdowns required to complete the project?

Additional power costs are not usually included as part of the project costs, but they should be identified to management, as they will affect the overall plant operating cost.

1.10   Instrumentation Requirements

Summarize what instrumentation work and/or equipment is included in the project. Check to see what section of the estimate your plant considers control computers to be under.

Some issues that should be addressed are:

- Are new computer systems required?
- Are new computers required?
- Are modifications needed to existing systems?
- Is a new network required?
- Does the plant use a double up system to ensure the control systems do not fail and if so how will the project affect this system?
- Will the plant air requirements be increased?

This is one area where a small addition can drive the direction of the plant for years to come. As an example, if you have to install a stand-alone computer system for process control and it is the first of its type and make in the plant, your selection will affect what product the plant will have to buy in the future. If your plant does not have a part standard and you buy Brand X for the project, then buying Brand Y at a later date may not work, as the two brands are not usually compatible. This then forces the plant always to buy Brand X even though it is inferior to Brand Y. If this is the case, this fact should be pointed out to management when completing the Project Authorization Document.

1.11   Safety

Summarize any safety concerns taken into account and what measures will be taken to address them during the project. Be proactive in your safety concerns. Depending on the plant, protecting workers can become a costly item. Some required safety equipment is specific to the plant or industry. Because of the high cost of such safety equipment, not all contractors or individuals will have the equipment. It may be cheaper for the plant or company to supply the special safety equipment, in which case your project will be charged for it. (Examples would be respirators and special scaffolding.)

Case History 4

When commuting to work on the North Slope in Alaska, we were required to carry winter survival gear during September to May. This survival gear was expensive, specialized gear and was supplied by the company we worked for. We checked arctic parkas out of the stores department when we left town and had to returned them when we got back. When on the North Slope everyone leaving the compound for one of the facilities had to carry a duffel bag of additional survival clothing. Our company had several of these duffel bags of gear for our use. The cost of this gear was written off over several projects. If new gear had to be purchased, the project had to pay for it.

1.12   Training, Commissioning, & Startup

Summarize what is required to train employees, commission, and start up the equipment. (See Chapter 13 for information on the difference between startup and commissioning). Will someone be brought in as a Startup Manager? At what cost to the project? What type of training is involved? Who will do the training? Where will the training be done? Will vendor's startup personnel be needed? How much lead time do you need to hire new employees so that they are trained before you start up the new equipment?

1.13   Permits

Summarize what permits you have included. Environmental permits can have a big impact on the plant and they should be highlighted in this area, if they apply. If building permits apply, explain any arrangements required, as sometimes the total cost is not known. Be aware that building permits only apply to the building structure and not to the mechanical equipment inside.

Case History 5

We were building a greenfield plant and had to get a building permit for the building only. When we applied for the building permit from the local municipality, we had to provide drawings that were available at the time and the permit cost was based on these preliminary drawings and on an estimated building value. When the project was complete, we had to provide the Building Department with "As Built" drawings and the final value of the structures. Additional money had to be paid to bring the building permit up to date with what was constructed.

1.14   Environmental

Summarize any environmental concerns that may arise from the project either during construction or after startup. Does the project affect any existing environmental permits? Are new permits required?

Estimating

In order to complete the Project Authorization document competently, you will have to prepare an estimate. Following are some of the factors that affect the accuracy of any estimate you prepare:

- The time available to prepare the estimate;
- The budget available to prepare the estimate
- Who is preparing the estimate
- Amount of up-front engineering done

When asked to prepare an estimate, one of the first questions should be "What is the required accuracy?" In terms of percentage, estimates can range from ±10% up to ±50% or greater accuracy. As stated previously, for a ±10% estimate the backup information you are using has to be detailed, current, and accurate. This requires a lot of up-front engineering work. In fact all the detailed engineering should be complete and current pricing obtained using these detailed drawings.

The pricing would include quotes from material suppliers, vendors, and contractors. This type of estimate requires the company to spend money up front to:

- dedicate a plant person to the project,
- have the detailed design done,
- prepare, issue, receive, and evaluate proposals, and
- prepare the estimate based on the above.

The ±10% estimate is the preferred type; however, as you can see a lot of time, money, and effort are required to complete this type of estimate. There are very few of these estimates.

At the midrange are the ±20% to ±30% estimates. For these you would have a preliminary flowsheet and layout and a basic scope of work. You could be using past equipment pricing with a multiplier. Based on the theory that prices keep rising, you multiply the past price by a value so that the new price is what you would expect to pay today. The multiplier allows for inflation plus other rising costs. (Your purchasing group can help you determine what multiplier to use.) You can also get a ballpark estimate price from a vendor or contractor (with their own accuracy range).

You will have looked at the installation requirements and calculated a lump sum estimate for them. You can get installation information from published man-hour summary books. There are several on the market; however, you should check with your purchasing group and contractors for confirmation of the numbers before you use them. If the directory says it takes 5 man hours to install a certain size motor, you should check with a contractor to see if it is realistic. The less information you have, the more your estimate will lean toward ±30% and the more information you have the closer your estimate will be to 20%.

At the end of the scale is ±50%. This estimate is based on very little information and a lot of experience. This would take very little time to prepare. The less experience you have, the more the estimate will be off and a range value higher than ±50% may have to be used.

Most consulting engineering firms have charts showing the relationship between project design and scope information available and the accuracy of the estimate.

To do the estimate properly, each job should be broken down into its basic components using the WBS. This will help ensure that all components of the work to be done are covered. It pays to talk to vendors, contractors, and plant personnel to determine:

- how the equipment goes together,
- how things will be constructed,
- manpower required,
- length of time to install,
- utilities required,
- equipment and special tools required.

Most plants have estimating forms and these should be used for the estimate. This ensures that the information is in a consistent format throughout and makes for easier understanding at a later date. This could prove to be important. The forms have several columns with the following headings:

- Item
- Quantity
- Unit
- Description
- Material
- Labor
- Total

Experience has shown that the more time spent up front doing the estimate, the more detailed and accurate the estimate will be. The estimate accuracy will determine the amount of contingency allowed for, i.e., an accuracy of ±50% will require more contingency money than an accuracy of ±20%.

Because of the time it takes to get projects approved, it is important to put as much information as possible in the estimate. The information can be on the estimate forms, in writing, or on sketches/drawings. Unless you have a very good memory, it is very difficult to remember specific items later on.

As you get into the estimate, questions will be raised about different items being included. Unless you have cross referenced items, it may be difficult to determine if specific items have been included. You may place an item in a section because at that particular moment it made sense to you to put it there. However, later on you may wonder why. For example, if you estimate the cost of piping including insulation, you would note in the Piping Section that the cost includes pipe plus insulation. In the Insulation Section you would make a note that the insulation is covered in the Piping Section.

You can see what advantage this cross referencing is. You now know that pipe insulation has been covered and where the estimate is located. Anyone else who looks at the estimate can see this as well. This will make finding things very easy and help ensure that everything is covered.

As each of the estimate sections is prepared, items forgotten or not thought of will be brought forward to be included or excluded. The important point being that the item is identified. Those items excluded should be noted so you know what has been looked at and do not have to cover the same ground twice.

The estimate should be made using the Plant Project Accounting Cost Code system. Then when the project is approved, the estimate can be given to accounting for direct entry into the cost reporting system. Do this up front as you go, as you do not want to revisit the estimate to put the cost codes together.

When doing projects with outside consultants, the plant cost code numbers should be used since the plant personnel will be familiar with them. There may come a time several years down the road when the estimate will have to be referred to and having an understanding of how the information was compiled and listed will help in the retrieval of information.

Engineering design hours, draftsman's hours, etc. that are to be charged to the project should be shown as a line item in each section of the estimate. This can be a bit fuzzy, as some plant personnel do not charge their hours to a particular project while others are required to. You should check first with accounting to see how your plant handles these costs. If an outside consultant is going to do the design, his or her design man hours and costs should be shown in each section if possible.

As the project engineer you will:

- work on the items you are capable of doing,
- coordinate the collection of information from others,
- compile the information,
- write the final document.

When preparing to do an estimate, your accounting group will have a list of standard headings for the estimate. These headings are all the areas that should be looked at for a complete estimate. These headings tie back to the plant accounting cost codes. These lists are fairly standard throughout industries in North America, the main difference being the numbering sequence. Your plant list may only be suitable for your plant projects, as very large projects will have a more detailed list of items. Below is a fairly comprehensive list of items that should be included in scopes and estimates. Not all the items apply to very small projects, but by including as many items as possible, the documentation will be more complete and professional. As well, by being disciplined in using the list for small projects, you will be better prepared for larger ones.

For estimates that have to be done quickly, it is extremely important that any and all assumptions are spelled out and the items that are not included in the estimate are listed as exclusions. The following list is more than is required for small projects. You can add items as necessary for larger ones. By using the applicable items from this list, the estimate will be fairly complete. The numbering refers to the Authorization Checklist in Figure 2.4.

Authorization Checklist		
1.0	**SCOPE**	2.11.2.5	= Single Line Diagram
1.1	Project Objective	2.12	Electrical Equipment
1.2	Schedule	2.13	Electrical Installation
1.3	Critical Assumptions	2.14	Electrical Tie-ins
1.4	Drawings	2.15	Instrumentation
1.5	Standards & Specifications	2.15.1	= Control
1.6	Process Requirements	2.15.2	= Control Loops
1.7	Battery Limits	2.15.3	= Control Panels
1.8	Mechanical/Civil/Structural Requirements	2.15.4	= Control Room Equipment
1.9	Electrical Requirements	2.15.5	= Field Instruments
1.10	Instrument Requirements	2.15.6	= Instrument Installation
1.11	Safety	2.16	Air Purification
1.12	Startup & Commissioning	2.17	Environmental
1.13	Permits	2.17.1	= General
1.14	Environmental	2.17.2	= Regulations
2.0	**ESTIMATING**	2.17.3	= Permits
2.1	Site Work	2.17.4	= Working Conditions
2.1.1	= Demolition & Relocation	2.17.5	= Surrounding Community
2.1.2	= Surveying	2.18	Training
2.2	Site Grade	2.19	Startup
2.3	Site Improvements	2.20	Permits
2.4	Structures	2.21	Engineering
2.4.1	= New Structures	2.22	Construction
2.4.2	= Existing Structures	2.22.1	= Inspection
2.4.2.1	= Modifications	2.23	Process Simulation Trials
2.4.2.2	= Additions	2.24	Insurance
2.5	Special Access Requirements	2.25	Currency
2.6	Soils	2.26	Contingencies
2.7	Mechanical Equipment	2.27	Escalation
2.7.1	= Process Equipment	2.28	Freight
2.7.1	= Mobile Equipment	2.29	Restrictions
2.7.3	= Other Equipment	2.30	Work By Others
2.8	Piping Requirements	2.31	Services For Construction
2.8.1	= Pipe Testing & Inspection	2.32	Utilities For Construction
2.8.2	= Piping Tie-ins	2.33	Receiving & Storage
2.8.3	= Fire Protection	2.34	Safety
2.8.4	= Pipe Insulation	2.34.1	= Design Considerations
2.8.5	= Pipe Penetrations	2.34.2	= Plant Hazards To Contractors
2.9	Other Insulation	2.34.3	=Construction Hazards To Mill Employees
2.10	H & V	2.34.4	Safety Issues After Construction
2.11	Electrical	**3.0**	**SCHEDULES**
2.11.1	= Power Source	3.1	Spending Forecast
2.11.2	= Power Distribution	3.2	Overall Project Schedule
2.11.2.1	= General Distribution	3.2.1	= Engineering
2.11.2.2	= Item Distribution	3.2.2	= Procurement
2.11.2.3	= Wiring & Control	3.2.3	= Construction
2.11.2.4	= Grounding		

Figure 2.4 Authorization Checklist

2.1 Site Work

2.1.1 Demolition & Relocation

Does anything need to be demolished or relocated to prepare the site for the project? This includes structures, equipment, piping, electrical, underground utilities, instrumentation, etc. Be sure to indicate what will be done with any equipment that is removed; i.e., reinstalled in another project, stored for future use, scrapped, sold, or whatever. If money is to be recovered, this should appear in the estimate. Make sure you are not overly optimistic in its worth. Also, be sure to identify and cross reference any rework or modifications required as a result of demolition and relocation.

2.1.2 Surveying

Is any surveying required to lay out the site, establish benchmarks, carry out legal surveys, or make up plot plans and submit to local authorities? Are survey points readily available on site or does a surveyor have to find them? Equipment location and setting of equipment should be under Mechanical Installation. If you have survey points that are currently outside, they may have to be moved inside if a building goes up and obscures the points. You should make an effort to protect survey points for future use. The plant should hire a surveyor to bring the required survey points and bench marks into the construction area for the contractor's use. Do not depend on your contractor to do this. You will have to bring the surveyor back throughout the construction to check certain points to ensure that the contractor is in the right location.

2.2 Site Grade

Are there any special site requirements such as clearing, excavation, fill, special elevations, slopes, surcharging, etc.? How will waste material be disposed of? Is there a pollution problem with any of the areas to be graded? If there is a need for a soil engineering study, identify this and cross reference to the appropriate sections. Does a topographic survey have to be done? Will winter work be a problem?

2.3 Site Improvements

Are there any new requirements for, or modifications to, existing roadways, railways, area paving, area lighting, underground utilities, landscaping, parking lots, fencing or other security measures, etc.?

2.4 Structures

2.4.1 New Structures

Does the project require new structures such as a new equipment building, control room, electrical room, offices, etc.? Include any requirements for foundations, floors, walls, ceilings, roof, painting, linings, lighting, heating and ventilation, fire protection, etc. Be sure to identify items such as restrooms (male and female), locker rooms (male and female), laboratories, maintenance shops, offices, electrical rooms, etc. Are there requirements for new building structural steel, such as towers or pipe supports? (Platforms and equipment supports should be included under mechanical equipment.)

Allow for freight if needed and cross reference to the freight section.

2.4.2 Existing Structures

2.4.2.1 Modifications

Does the existing structure have to be reinforced, walls moved, painted, etc.?

2.4.2.2 Additions

Are there additions to be made to an existing structure? If you add an extension to a tower in this section, you will have to add the reinforcing of the tower in section 2.4.2.1 and cross reference.

2.5   Special Access Requirements

Most building codes now require special walkways for the handicapped as well as handicap accessible washrooms; in fact, no new addition or building should be constructed without considering handicap access. Depending on your jurisdiction, you may be required to install handicap washrooms in all areas but you do not necessarily have to provide handicap access to the washrooms. It is important to check this out. Requirements can affect how you approach the layout and the design. As an example, if you plan on a two-story building but have to put in an elevator for handicap reasons, it may be less expensive to go with a single story. This can be a costly item if not checked out properly at the estimating stage.

Are there any out of the ordinary ladders, platforms, doors, lighting, etc. required for access to facilities and instrumentation for their operation and maintenance? Can items be reached from manlifts or cranes, etc? One way or another everything in the plant should be accessible. If you plan on using existing plant equipment to access items, you should state that here, e.g., a manlift will be used to access valve handle.

2.6   Soils

Identify work to be done or work previously done to determine the soil conditions for the foundation design. Keep in mind it can be dangerous to try extrapolating existing soil information to cover an area that was not tested. You could run into major construction problems if you are wrong. For important items it is prudent to test, although testing is no guarantee either. Allow enough money to do a proper soils examination, i.e., do not use the local water well driller when a proper soils examination should be done.

If you are working inside a plant, check photos from the original construction or, if possible, talk to people who worked on the construction to get a feel for what is under the building. If need be dig test pits inside the building, but try to do something to make an educated guess as to what your foundation requirements will be.

Case History 6

For one project we were trying to determine the depth of rock in an area where a new building was going to be constructed. We knew there was rock at a shallow depth and that it sloped upwards across the site. We did borings every 2 ft. across the width of the site. The borings indicated rock at a depth of 15 ft. on the east side and at a depth of 2 ft. on the west side. The building foundation was designed accordingly. When we started excavation for the foundations we found that on the west side the rock dropped off 20 ft. We had stopped our borings 2 ft. from the drop off! Work had to be stopped while the foundations were redesigned. This affected the schedule and the project cost.

2.7   Mechanical Equipment

2.7.1   Process Equipment

For each piece of equipment such as pumps, conveyors, agitators, etc., include the following information:

- Equipment specification (e.g., pump type and model, agitator model, etc.)
- Materials of construction (e.g., 316 SS, titanium, etc.)
- Drive requirements (e.g., adjustable speed drive, motor HP, gearbox model, etc.)
- Special requirements (e.g., mechanical seals, slide bases, etc.)
- Ancillary requirements (e.g., flushing water, seal water, etc.)
- Spare parts
- Freight allowance, FOB, DDU, duty, taxes, etc. Mention what it is and cross reference.

If you have a quote from a vendor, include the quote number and date. Identify any other documentation that supports the price, e.g., fax, record of discussion, minutes of meeting, etc. File these directly with your estimate documents.

Installation should be included in this section. If there are any special installation requirements for the equipment, identify what is beyond normal practice and either cross reference or included costs, e.g., special cranes to do the lifts, mill conditions and how they will affect installation productivity, etc. Is an equipment installation surveyor required? Mechanical equipment surveying is a specialized field and it is wise to use an experienced equipment installation surveyor.

2.7.2   Mobile Equipment

Mobile equipment should be described in great detail. Most equipment is purchased for a specific application peculiar to the industry and there are usually small details in the equipment that the industry requires. Spare parts should be included. If there is too much information to write out, attach a copy of the equipment specification to the back of the estimate.

If you are buying a used piece of equipment, the whole issue of rebuilding and available spare parts has to be looked at in great detail. Field trips may be required to look at the equipment. You may want to hire an expert to look at it for you. His costs should be included in the estimate.

For both new and used equipment, record the vendor's quotation number and date as well as any other documentation that supports the price, e.g., fax, record of discussion, minutes of meeting, etc.

Allow for freight, duty, and taxes, or note how such items are being handled and cross reference.

2.7.3   Other Equipment

This would include maintenance equipment, laboratory equipment, test equipment, office equipment, etc. Maintenance equipment would include not only such things as lathes, welding machines, special tools, etc., but also cranes, hoists, monorails, etc. needed to maintain installed equipment.

Allow for freight or note how such items are being handled and cross reference.

2.8   Piping Requirements

The piping should be broken down to each individual line if possible. (This will depend on project size.) This breakdown can be done by pipe size, line number if available, material specification, or service, e.g., all 6" diameter pipe or all 316 SS pipe or mill waterpiping (all sizes). Describe and allow for any special requirements such as pickling, polishing, lining, flushing, etc. An allowance should be made for fittings, typically 10%. Required valves should be listed by size, type, and materials of construction. Large motorized valves should be listed as process equipment. Pipe supports are part of the piping; however, large structural supports will be part of Section 2.4.1 New Structure. For steam piping and other piping that will generate a high load on the building steel, a structural engineer should look at the loads to make sure the steel will support the loads. Pipe stress analysis may be required for some of the lines. If you have pipe standards for the piping, attach them to the estimate.

If a consultant is doing the piping drawings, you should agree on what size of pipe is considered small bore. Small bore pipe is field run and isometrics are not made up for them. Small bore is usually 2" diameter and smaller. For large bore piping, will pipe spool drawings be generated and by whom? Who will make up isometrics? At what cost?

2.8.1   Pipe Testing & Inspection

Is there a need for pressure testing of the piping? Non-destructive testing (NDT), such as x-raying of the welds and use of dye penetrant, depends on what code the pipe is designed to. X-raying of piping has to be done after hours or the area has to be cleared of people and costs will have to be accounted for. Specialist will have to be brought in to do this testing. What code is the pipe designed and constructed to? This should be identified in the estimate. All pipefitter welders have to be tested and approved for the type of welding to be used. This weld testing of pipefitters is not part of the plant costs.

The owner should hire any inspection agency, so this should be covered in your estimate. Do not have the contractor hire the inspector. You want an independent analysis of the work and the inspector owes the information to the person he is contracted to. Are there any permit costs associated with the inspections? Who will prepare any permit applications?

In most jurisdictions pressure piping has to be sent to a government body for review and approval. There will be a time delay associated with this process so allow for this in the schedule and add the cost of the review in the appropriate section.

2.8.2   Piping Tie-Ins

Include any major tie-ins that will be required, giving location, tie-in number (if assigned), time restrictions, special requirements, etc. Be sure to look at fresh water, compressed air, steam, condensate, process liquids, vents, sewers, etc. In existing plants, do walkways have to be covered or barricaded to do the tie-ins? Does traffic have to be disrupted? Look at access to the tie-in location and allow for scaffolding, manlifts, etc. to get tradesmen to the point in question. Does cabling or instrument air have to be moved to allow the tie-in to be made? Do they have to be extended, relocated, or new cables run?

If tie-ins are going to be done without shutting down the line (on the run), how will the mill conditions affect the installation crews? Will work crews have to wear gas masks? Will they have to work in protective clothing on a hot day? Work conditions that are out of the norm like these should be discussed with the contractor and costs for lost production noted.

2.8.3   Fire Protection

Do changes have to be made to the existing mill system? Do additional sprinkler heads and sprinkler piping have to be run? Changes should be discussed with the plant risk department. Be aware that changes to the system have to be arranged well in advance and that there will be restrictions as to when the modifications or tie-ins can be made. Tie-ins may have to be done on a down day when no one else is around. The risk department usually has to notify the insurance underwriters any time changes are made to the system or parts of the system are taken out of service. Additional insurance may have to be purchased. It is prudent to keep the risk department advised of your intentions.

Do you have to get Fire Marshal approval? Can you discuss the project with him during the estimate stage to find out what he requires? Will his approval affect the schedule?

Case History 7

We were installing equipment for a project in the US that used a product called Thermal Oil. It is heated to high temperature in boilers and under certain conditions can create a fire hazard. We had to get fire marshal approval on the boilers and piping system. Unfortunately, the fire marshal we had to work with was the same one who had approved a similar system that blew up and created the worst industrial fire in the state. We were not aware of this when we started out, and there were delays in getting approvals and extra money had to be spent with consultants to get the necessary approvals. This was never accounted for in the original estimate.

If you have to add fire extinguishers, the costs may be charged back to the project. Check the company policy on who supplies them. The same goes for hose reels. Is the fire pump adequate to cover the new addition? Are any existing permits affected by the changes? Firestop around electrical cables, cable penetrations, and pipe penetrations should be included here. Your electrical group can give you advice on what the products to use.

2.8.4   Piping Insulation

Insulation should include any special requirements giving type, extent, lagging, jacketing, thickness, etc. Are flange shields required? Tie-in repairs should be included as well as lengths of pipe. As with pipe, insulation can be grouped by line size or material carried (e.g., steam) or insulation material.

2.8.5   Pipe Penetrations

Where the pipes go through the floor, walls, or roofs, the openings have to be sealed and flashed. For electrical penetrations there is usually a firestop foam material installed around the electrical cables and tray. It works equally well with piping through walls. If the sealant is to be used as a firestop it should be included in Section 2.8.3. If you require a material to keep weather, animals, and insects out, it should be accounted for here. All pipe, ducting, and electrical penetrations should be sealed, especially if going outside. Sometimes the penetrations will go between a heated and unheated area and other times into a sealed area. These types of penetrations should be sealed. Your plant risk department should be able to tell you which areas must have a firestop for penetrations. (Halon should discharge into a closed area.) Roof penetrations are potential leak problems and should be avoided if possible. If you have to go through the roof with pipe, these penetrations should be sealed with a coned shaped rubber boot that fits over the pipe and is sealed to the roof deck.

2.9   Other Insulation

This includes other major insulation requirements such as tank insulation, equipment insulation, noise abatement insulation, insulation for personnel protection, etc. Personnel protection is insulation installed strictly to protect personnel and it has no process requirement. An example is insulating a section of a hot pipe only in the area where a person is likely to touch it. This section should be cross referenced with the appropriate equipment section to make sure the insulation for each piece is covered.

2.10   H & V (Heating and Ventilation)

Beside including building heating, makeup air units, and exhaust fans, this section should also include ventilation for drive motors, air conditioning for Motor Control Center (MCC) rooms and offices, etc. Keep in mind that as more air is taken out of the buildings it has to be replaced; otherwise, the building pressure becomes negative creating other problems. Make sure the building can handle additional loads if new equipment is to be placed on the roof.

2.11   Electrical

2.11.1 Power Source

This section is the high voltage area and should cover new substations, incoming cable feeds, substation breakers, etc. —basically anything to do with the incoming main feed to the plant. Decide whether the feeders to the MCCs are included in this section or under Power Distribution. Either way, it should be noted so it is not overlooked. Any costs associated with the local power utility should be indicated here.

If new transformers are required, all the civil work associated with it should be included. As with equipment, an allowance for freight should be made here and cross referenced.

Will the project require an increase in the main plant power supply? Will the project put the plant into the next power block? Are negotiations required with the local power authority?

2.11.2 Power Distribution

This usually includes everything from the MCCs to the equipment, including the MCCs. Any Programmable Logic Controllers (PLCs) associated with the MCCs should be in this section. Below are areas that should be looked at as part of Power Distribution.

2.11.2.1 General Distribution

Some distribution areas in the plant are not identifiable and any revisions or additions to these areas should be listed under this heading. Include the major equipment required and location, any special features, and freight. Cross reference as required.

2.11.2.2 Item Distribution

This covers the power distribution to each piece of equipment and includes the major equipment required and location, any special features, and freight. Note whether or not a new main feed from the plant substation is required and if so what section of the estimate it is in.

2.11.2.3 Wiring & Control

Following is a list of the wiring and controls that could be required for each piece of equipment. Include what equipment is required as well as installation:

- Medium voltage control centers
- Low voltage control centers
- Special controls (variable speed, soft start)
- Instrument wiring
- Special services wiring

2.11.2.4 Grounding

Include all grounding and installation costs for the power systems, equipment, switchyards, computers, distributed control systems, etc. Check the size of the ground grid as some ground grids can be quite large and could interfere with equipment layouts as well as transformer/substation compound sizes. In new construction there will usually be an underground ground cable that will be attached to the building steel and some equipment.

2.11.2.5 Single Line Diagram

A single line diagram should be prepared to show how the new project ties in with the existing plant electrical system. This diagram shows all the electrical equipment from the incoming feed to the motors. Who will prepare the single line? What is required for discussions with the power authority? This drawing is required in order to have a discussion with electrical contractors so they understand what is required and to get a reasonable estimate.

2.12   Electrical Equipment

Using the following list, include costs for major items of electrical equipment including installation:

- Electrical equipment requirements
- Equipment specification
- Special requirements
- Ancillary requirements
- Spare requirements

Remember to cross reference to other areas carrying items for electrical equipment, e.g., ventilation, air quality control, etc. Look at the size of the electrical equipment and make sure it can be installed without building modifications and note what lifting equipment is required to place it. Don't forget to allow for freight and cross reference.

2.13   Electrical Installation

Include for installation of items such as raceway and conduit, cable, motor connections, etc. Watch for congested areas, overhead work, scaffolding or manlift requirements, etc. Don't forget firestop where cables pass through floors or walls. If installing cabling underground, you will have to consider how to schedule it with other construction. Don't forget about traffic flow if crossing roads. With most electrical installations shutdowns in the off hours are required to get new electrical sources transferred.

2.14   Electrical Tie-Ins

Include for electrical tie-ins, giving location, time restriction, special requirements, etc. If the design is far enough along, include tie-in numbers as well. If major plant shutdowns are required they may have to be done in the early hours or at some other non-standard time, in which case other contractors may not be allowed to work, etc.

2.15   Instrumentation

2.15.1 Control

What type of control is required in the plant—Distributed Control System (DCS)? PLC? Is the new control part of a larger plan for an overall upgrade to the existing plant system? If so this should be mentioned since you may be installing a more powerful system now than is required and this will be reflected in the project cost. What you do in this section may set the direction for other projects and this information should be common knowledge in the plant. You do not want to spend money on technology today that will be thrown out in a few years as improvements are made, unless that is what management wants.

This section should also include the hardware and software costs. All equipment listed should have specifications included. Installation should be allowed for. You should use an Input/Output (I/O) count to estimate the termination costs.

2.15.2 Control Loops

Control loop drawings have to be prepared for proper installation. They are the road map by which the wiring is hooked up. Will they be done in-house or by an outside consultant? Is a control architecture drawing required? Cabling for all controls should be included here. Installation should be the pulling of cable only. Terminations should be with the equipment installation. Keep in mind that some instrument cable has to be isolated and can not be run in electrical trays. Make sure you get control loop drawings from all your vendors. This requirement should be in the purchase order documents.

2.15.3 Control Panels

Are new control panels required or can existing panels be added to? Are panels being added in the field or just in the control room? Field panels may require mill air to keep the environment clean. If panels are bottom fed, holes in the floor may be required, etc. Are weathertight or explosion-proof enclosures required? What is the metallurgy of the panels to be? Do they need a protective cover because of the atmosphere they are in? Are paint specifications required? Include the panel installation and cable termination.

2.15.4 Control Room Equipment

If a new control room is being installed all furnishings, panel cabinets, computer floor, lighting, etc. should be included in the estimate. New lighting may include anti-reflective fixtures. In high noise areas soundproofing may be required.

Do not forget fire protection, either new protection or upgrades to the existing protection. The existing control room may be protected by Halon, which is no longer manufactured. Existing supplies are getting more expensive, so alternatives will probably have to be considered. Alternatives include carbon dioxide and water mist, to name a few. The choice you make today could set the direction the company will have to follow in the future. This decision should be pointed out to management.

2.15.5 Field Instruments

Using the following list identify and include for field instruments such as flow meters, temperature transmitters, control valves, etc.

- Instrument specifications
- Ancillary requirements
- Spare requirements for computer boards—this is usually a percentage of the number and type installed.

2.15.6 Instrument Installation

Include for the installation of field instruments and materials required such as tubing (stainless, plastic, etc.), fittings, raceways, heat tracing, cable, etc. Allow for terminations at the instrument. Are there to be redundant loops in case of failure of a loop? Don't forget fireproofing when going between floors or buildings. A lot of instrumentation is located in hard to reach places and requires equipment to get to it. Include for access during construction and look at what is needed for permanent access.

2.16 Air Purification

Are there any air purification, heating, and/or cooling requirements for control rooms, panels, and rack rooms? Does the existing system have to be upgraded? Is the new equipment compatible with existing equipment? Will smell be a problem? How about particle size?

2.17 Environmental

2.17.1 General

This would be items that have to be installed for environmental reasons to meet regulations and permits. The plant environmental specialist should review any project items that will affect existing permits and regulations in any way. Be careful that existing regulations and permits are not violated as this may allow existing permits to become open to scrutiny by people opposed to your operations. If this is the case with this particular project, this information should be highlighted and brought to management's attention in advance.

Be up-front in your dealings with environmental agencies; otherwise, they can make it a difficult and time-consuming process to get approvals and your project could be delayed. Although you may look on the environmental aspect as a project annoyance, it can have disastrous affects if not handled properly.

Items to review:

- Air emissions
- Effluent
- Solid waste
- Toxic substances
- Industrial hygiene
- Radiation
- Noise
- Hot and cold drafts

When you add onto a plant or construct a greenfield plant, you will get complaints from the local long-time residents that the plant or new equipment has changed the weather patterns and is now the cause for all the troubles in the area. It is important to consider environmental modeling and baseline testing to prepare for this in advance. The modeling will show what the weather and wind patterns are in the area and how any emissions from the plant will be dispersed. Baseline testing involves measuring the existing conditions at selected points in the area before changes are made. After plant startup the sites are monitored on a continuous basis to determine any affects from the plant. This information can be used to deflect criticism.

2.17.2 Regulations

What work and/or equipment are required for the plant to comply with existing regulations? Consider future regulations that are currently being reviewed or that will be implemented in the near future. Will these be covered by the work of this project? You should spell out what regulations are being complied with.

2.17.3 Permits

What work and/or equipment are required for the plant to meet its existing environmental permits? Will any work on the project result in the existing permit being exceeded? If so what is being allowed for in the project? If permit changes are required, the schedule may have to reflect permit approval time. Are the permits for construction or operation? The permits should be in place at startup. Is baseline testing required to determine existing conditions? If so, get estimates and include the costs.

2.17.4 Working Conditions

What work and/or equipment are required to limit the effect on working conditions? If noise from new equipment is going to affect existing areas where people are working, what will be done to limit the affect? Do people have to work in the cold or heat for extended periods? If so, how do you protect them? How will communications be handled in a noisy environment?

2.17.5 Surrounding Community

What work and/or equipment are required to limit the effect on the surrounding community? If noise from new equipment will bother the surrounding community, what is the project doing to limit the affect? Noise, odor, fumes, and particulate matter are usually the problems that have to be solved. Will a community hearing be required? Will consultants need to be hired?

2.18   Training

Most plants have separate budgets for training, but this should be checked so that money is not left out of the estimate. The size and type of equipment will determine the extent and quality of the operating and maintenance training required. Check with equipment vendors as some will provide training as part of their startup package and the costs can be included with the equipment. If you have to include training, the following items should be considered:

Scope of Training

Identify the number of people to be trained and the skills needed. What is the length of training time for each position? Are the people to be trained new hires or existing employees? Do operations personnel have to come from another plant for training?

Materials and Facilities

Will special manuals, video programs, or outside courses be required? Will special classrooms, labs, or prototype equipment be needed. Are trips to other plants or manufacturers' plants required? Are additional computers required? You should try to get as much of this material in the vendor's quote and included in the vendor's purchase order.

Training Personnel

Will training be done by plant operations employees, staff personnel, or outside consultants? If vendors are providing the training, is it in their quote?

Time on Machines

Identify the amount of time personnel will be training on the machines. Will they be classed as extra crew only for the purpose of training? If time is to be spent on the machines, where will this take place—at the plant or off site? This could take place in a foreign country for a period of up to three or four months.

2.19   Startup

Check with the accounting group to see how your particular plant handles startup costs. Commissioning is the running of the equipment as a complete unit without raw materials and is the point of completion for the contractor.) Commissioning costs are covered by the project. Startup is the running of the equipment as a complete unit using the actual raw materials. Startup involves Time & Material Contracts and is not a fixed time period. Check all purchase orders to see what startup costs have been included by the vendors in their price. If nothing is included in the vendor's price, then determine what the cost is? You should get a daily rate from the vendor for startup personnel for time spent on-site over and above what is allowed for in the vendor's quote. Vendors differ on how training and startup are handled. Sometimes the person doing the training will do the startup; with other vendors training is done by someone other than a startup person. Other times the vendor's sales representative will do the training and startup.

How long will the startup take? You have to set up guidelines to indicate the difference between commissioning and startup (see above)? Will plant personnel working on startup be charged to the project or are they covered by their department budgets? This information has to be known in order to come up with a reasonable estimate for startup costs. Depending on your process you may have to buy material from an outside source to commission your new equipment. For example, if you are going to manufacture plywood, you would have to purchase sheets of plywood from another source to set up and commission your saws.

Contractors' quotes will take you through commissioning but will not include startup. Startup is usually handled as a time and material contract. You can also hire the required tradesmen for a fixed time period, say three months. The only problem is you have to hire all trades but you may not have work for all of them so you are wasting money. It is better to do a time and material contract.

2.20   Permits

The cost of acquiring all project related permits should be included. This can include building permits, environmental permits, electrical permits, etc. Permit costs, such as a building permit, are based on the estimated final cost of the building structure. This cost is not accurate at the beginning of the project and therefore the initial permit cost is based on an estimate. At the end of the job the balance of the permit fee, based on the actual final cost of the building, is paid. Be careful—there can be a lot of unknown costs in this area.

Identify all permits and licenses required by outside agencies that will be needed for the project. This could include:

- building permits
- environmental permits
- electrical permits
- transportation permits
- police and fire department permits
- radio licenses for two-way radios
- government boilers branch approvals
- Department of Labor approvals
- fire marshal approvals
- environmental approvals including hearings

The government approvals can be costly and time consuming. Depending on your location you may have to hire an outside consultant who knows the regulations in the area. Some locations can have regulations you may not have thought of, so it is prudent to investigate at the approval stage. If you have not planned ahead, a permit delay can seriously affect your schedule.

Case History 8

We were fast-tracking the building of a greenfield plant in California. We were not aware of the building permit requirements when we started and ran into numerous delays as a result. When we poured concrete foundations, the building inspector had to come out and check that the rebar was installed according to the drawings. (We already had a civil site inspector doing that, so it was done twice.) He also had to inspect every third course of masonry block and count the number of drywall screws before the joints could be taped. In Canada this was unheard of. For the California site we had to hire a person who was familiar with the local building codes and who knew when to call the building inspector in. Once the inspector was called we then had to wait for him to show up since there were not that many available in the county. The building inspector problems caused unanticipated delays in the schedule.

Building drawings, even though stamped by a professional engineer, had to be reviewed by the Building Department. This caused a delay of several weeks, adding weeks to the schedule. We had to complete engineering drawings several weeks early to get them approved by the Building Department. Once they were approved we were not allowed to make changes, even though it was standard practice in the industrial field. More than once the building inspector shut down construction because the design engineer had made a change that had not been approved by the building inspector so the site did not conform to the drawings he had. These permit issues were very costly to the project.

2.21   Engineering

The cost of outside consultants should be included. The consultant should provide an estimated cost to do the engineering work. You should check with the accounting department to see if and how plant staff is charged to the project. The project engineer's cost may be included as well. You will have to find out if the cost of plant personnel includes their benefits, etc. For some plants the personnel cost is their straight time hourly rate; other plants add benefits. If you are dealing with back charges to vendors, the plant will have a rate for your time (which will include your benefits) to be used in the back charge. You should check this out. If consultants are coming in from a distance, you may have to pay living out allowance (LOA), car rental, hotel, and airfare home. These costs can add up fast, so make sure you have them covered.

If using an engineering consultant there will still be costs associated with some of the plant office staff, i.e., accounting, safety, drawing control, drawing review, secretarial, etc. The costs of preparing the estimate are usually charged back to the project, once the project is approved. You should show a line item for estimate preparation.

2.22　Construction

The size of the project will determine if you are assigned to the project for a specified time in a location remote from your current place of work. If that is the case, the following costs will have to be considered:

- Temporary facilities such as office space for the project staff. This could be trailers or space in a plant.
- Is a washcar (male and female washrooms) needed or can contractors use the plant facilities?
- If fiberglass work will be done, special facilities will be required.
- Is office space for vendors going to be supplied during startup and where will that space be?
- How many people will be in the field office?
- Signage will be required to direct people, for traffic control, and to meet safety needs.
- Is a field vehicle required? This is usually a pickup truck. In some jurisdictions this vehicle may have to be used for transporting injured workers.
- Office furniture is either rented or purchased. You may need a drawing storage cabinet. How will site drawings be reproduced?
- Office supplies such as pens, pencils, and paper.
- What field stamps (RECEIVED, COPY, OBSOLETE, etc.) will be required?
- Office equipment such as fax machines, computers, telephones, postage meters, etc.
- Who is providing security and first aid? Are existing plant forces to be used?
- Who provides the fire protection?
- Is a first aid room required? How will it be staffed? This depends on number of people on-site and the distance from a hospital.
- Are a security gate, security trailer, and fencing required? What about after-hours security?

Identify in this section the cost of any programs that relate specifically to safety. The costs may be for signs, special safety programs, newsletters, and safety incentives for the contractor or project personnel, etc.

2.22.1 Inspection

At various stages in the project you will require inspection services. This is a cost borne by the plant, as you do not want the contractor paying for this. The contractor should cover the cost of government inspectors only. Quality control such as soil testing, concrete testing, weld inspection, etc. should be covered by the project. Anything you want inspected should be paid for by you. This way the inspector reports to you and not the contractor. Government inspectors will report to you, because it is your plant. You should also have a surveyor who you can use to check important items, e.g., anchor bolts, setting of monuments, etc.

2.23　Process Simulation Trials

Do any material trials have to be run to prove that the equipment or process is correct? The material trials may be run at an equipment supplier's research and development establishment. Sample material may have to be sent from the plant and personnel may have to go to witness the trials. Sometimes equipment will be designed and sized according to the sample material sent.

Are computer process simulations going to be run? This usually involves a consultant at some cost to the project. If this work was done before the estimating and project approval stage, accounting will allocate the costs to a holding area and will charge those costs to the project upon approval. Find out if and what costs accounting has accrued for your project and include them.

2.24   Insurance

Although the plant has insurance and contractors carry their own insurance, additional plant insurance may be required for larger projects. The plant insurance group should be able to advise if this is required and supply a cost for the estimate.

2.25   Currency

If you are buying equipment from outside the country, be aware that the exchange rate can have an affect on your costs. Any quote you receive from a foreign vendor have a time limit on how long the quote is valid. Because equipment purchases are staggered, you can, at the time of equipment purchase, fix the equipment price by buying currencies forward. This will give you a fixed value for the dollar currency until all payments are made. If there are fluctuations in the currency, you may even make money on it.

2.26   Contingencies

A contingency covers unforeseen costs that may come up over the life of the project. Contingencies are usually a percentage of the estimated cost and are added to the estimate in each section. On small projects 10% is average but this will vary with the detail involved in the estimate. In areas where the estimate is based on a firm price the contingency will be less. Each plant will have its own percentage to use and this should be available from the plant accounting group. Realize of course that on a $15 million dollar estimate you may not get a $1.5 million dollar contingency.

Throughout the project guard your contingencies well as there is a tendency by others to use them to cover any conceivable item. If the request is a scope change, use the appropriate forms and procedures to get additional funds. If you start using contingencies for scope changes you will have nothing left at the end of the job when you really need it. In times of tight money the pressure will be on to add all kinds of items to the project, and you will have to learn how to say no in a diplomatic way. The easiest way is to ask for additional funds and let management say no.

2.27   Escalation

Escalation is an increase in your costs due to inflation or events beyond your control. Examples would be:

- increases in costs of labor (because of a new labor agreement)
- changes in foreign currencies before equipment is purchased
- a change in market conditions affecting the cost of equipment.

Discuss with the contractor the possibility of doing the work with no escalation in labor rates. If a labor contract ends during the life of the project, it is unlikely the contractor will not know what the cost effect will be. If the labor contract extends through the life of the project, the contractor will know if escalation will affect his price.

The plant purchasing group should be able to advise on a percentage value to include in the estimate. The longer the project runs, the more likely there will be escalation. Escalation does not apply as a straight percentage of the project total since some items in the estimate are fixed. All fixed costs or costs you are confident with should be subtracted from the total and the percentage increase applied only to the remainder. If unknown, add something based on a best guess. Under Critical Assumptions make a note of the escalation value used. Management may want to delete this to cut the cost of the project.

2.28   Freight

Each section of the estimate should have an item for freight. Make a note in the equipment sections if the quoted prices include freight as FOB (free on board). It is important to know whether or not freight is covered. This can be an expensive item, so don't forget to allow something even if it is a guess, but don't guess too low. Talk to your purchasing group and transport companies to get a feel for costs. When getting quotes from vendors you should get prices FOB plant site. Any equipment from outside the country should arrive with duty and taxes included delivered to the plant site and should be stated as such in the quote. Equipment that has to travel across the ocean should be quoted as packed for ocean shipment. Even if the equipment has to cross a stretch of water on a ferry it should be packed for ocean shipment as it can get knocked around considerably once on the ferry. Do not make deals on freight with a carrier just to include in the estimate, as the deal may no longer apply when the project is approved.

Be leery of offers to arrange transport on a vehicle that will be picking up something else and can pick up your item along the way. Backhauling on trucks is inexpensive and usually not a problem.

Case History 9

We were trying to get a used paper machine from England to our company in Canada. As it was a large company we had a transportation group that made arrangements for the machinery to be put on a ship stopping in England to pick up a load of newsprint. When the ship arrived in England the captain took one look at the crates, plus the rest of the shipment and said "no way." We eventually had to make other arrangements to have it shipped by a regular freighter at a much greater cost.

2.29   Restrictions

Try to identify all potential problems that could affect the construction phase of the project. There may be additional costs such as having to work after hours to do certain jobs or having to wear protective clothing that lowers productivity. Some things to consider include:

- Mill operations that can affect construction
- Construction requiring the protection of existing facilities
- Mill regulations and work permits
- Having to use a certain labor force
- Lack of access to areas of the plant that may require overtime to obtain access.

- Restricted areas that can affect the contractor's production. These could be areas that require special protection to do the work and access is restricted to certain personnel.
- Not using mill facilities can affect costs and production
- Health and safety requirements - some equipment may have to be provided by the plant (see separate section below on Safety).

The issuing of work permits can turn into a costly exercise if not handled right. On major shutdowns you could end up with 25 or more foremen lined up at 6:30 a.m. trying to get an operator to sign a work permit. This is a cost you will have to absorb.

Case History 10

We had to install a trench screen in the main plant effluent trench. The area was susceptible to H_2S gas. The screen had to be installed on the run so for safety reasons the contractors' personnel had to wear Scott Airpacks and rubber protective clothing during the installation. It was a time-consuming and costly operation.

2.30   Work By Others

During the construction phase you should consider the effect the following may have on the project:

- Other contractors who will be working in the area, union and non-union;
- Plant maintenance crews who may want to do some of the work;
- Work done off site by non-union shops may have an affect on site work.

2.31   Services For Construction

For the construction phase of a project you will have to investigate and include costs for the following, if they are needed:

- Construction parking
- Separate construction gate
- Signage
- Location of contractor buildings such as offices and change shacks
- Inside storage - can exiting mill areas be used?
- Receiving - will the plant receiving area and personnel be used?
- Outside storage areas
- Shake out and assembly areas
- Security requirements - will any be provided? Will existing plant security cover the job site around the clock? Will there be security at the gate leading to the site?
- Changing of plant traffic patterns
- Fire watch and fire department
- First aid and safety
- Camp costs

2.32   Utilities For Construction

The following should be looked at to see what costs should be allowed for if the utilities are not available on site:

- Electricity/construction power – you may need a new transformer
- Construction power panels
- Washroom facilities, both male and female
- Mill water
- Portable water
- Temporary lighting
- Compressed air
- Telephones, communications
- Waste disposal (hazardous waste disposal is handled by the contractor)

2.33   Receiving & Storage

Receiving

For small projects the plant stores can receive the project materials. For large projects a contractor may do this or a special person may be hired to fill the role of receiver. If plant stores is doing the receiving, make sure your purchase order numbers are different than the plant's so your material can be readily identified and you can be notified when something comes in. You may have to include some money for stores and purchasing personnel to handle these issues.

Storing

On small projects materials can usually be stored in the plant warehouse. On larger projects additional storage and stores personnel may be required. Secure storage may also be required for some materials (e.g., instrumentation) and should not be overlooked. For rotating equipment in long-term storage someone has to rotate the equipment on a regular basis and costs should be included. (Pumps, fans, etc. have to be rotated to prevent flat spots in the bearings.) For certain items, inside storage is a necessity and all equipment vendors should be questioned about this. In places where snow is a problem, materials will have to be put up on timbers to keep the material out of the snow and make it easy to find. Consider your situation and allow money for storing equipment appropriately.

2.34   Safety

Carefully consider this area in advance as unexpected costs can arise from unanticipated problems. Safety issues should be discussed with the plant safety officer. If the contractors are to provide safety personnel and first aid they should know this at the time of tendering. What are the plant safety requirements? These are specific to each site and some requirements can affect the contractor's costs. What safety equipment will the plant supply (such as respirators)? Will they be charged to the project? It is usually not reasonable to expect the contractor to supply this. In most plants there is usually an orientation for new people coming onto the site. Advise the contractor how long this orientation process takes, as this will have to be included in the contractor's costs.

Case History 11

In order to work on the North Slope in Alaska, everyone had to take and pass a two-day safety course held at a hotel in Anchorage. This allowed one to get onto the North Slope. Once on the North Slope, a further one-day facility-specific safety and orientation course was required. These costs were charged back to the owner.

2.34.1 Design Considerations

Include costs for any items installed solely for personnel protection such as insulation, access ladders, platforms, alarms, special guards, etc.

2.34.2 Plant Hazards to Contractors

Include costs associated with protecting contractors from specific hazards they might be exposed to but may not be aware of. Examples would be unexpected upset conditions causing flooding, hot water, steam, chemical or gas discharges, other contractors, plant maintenance crews, plant operating activities, etc. If no cost is allowed, state so in your critical assumptions.

Do not include costs for plant evacuations as this will come out of your contingency. You will be charged for lost time due to:

- any plant evacuations, (e.g., fire, chemical spills, gas leaks)
- shutdowns on orders from government agencies (e.g., shutdown due to a serious injury or fatality on job site)
- fire drills.

2.34.3 Construction Hazards to Mill Employees

Include costs associated with protecting plant employees from hazards the contractors would be imposing on them, such as:

- overhead loads
- fire
- increased traffic
- noxious odors
- noise

Special walkways or traffic control may be required.

2.34.4 Safety Issues After Construction Completed

Include costs associated with safety hazards presented by operating or maintaining the new system. This would identify training needs. Also include safety hazards the new installation will impose on the plant in general. Are traffic patterns affected? Are new hazardous materials involved (gas, etc.)? If so, are Material Safety Data Sheets (MSDS) involved? Are new evacuation plans needed, etc.? How do you inform all plant employees? How do you inform all contractors/visitors who come on site infrequently?

3.0   Schedules

Two types of schedules should be included with the project approval documents—the spending forecast or cash flow and the overall project schedule.

3.1   Spending Forecast

This is a forecast of when project money will be spent. It is used by the financial group to determine how much cash to have on hand and when. They will combine your spending forecast with other spending forecasts to get the total cash required at various times throughout the project life. For project approval, the spending forecast is your best guess estimate and should include an allowance for project approval time.

Once the project has been approved forecasting will become a routine exercise, at least monthly. The accounting group requests the forecast and will usually send you a standard form to fill out. Remember that with contractors the spending starts shortly after award of a contract, as they will want money for mobilization and commitments they make for material. Some material may be in the contractor's shop and will have to be paid for before it shows up on site. Be aware that construction contracts usually stipulate progress payments and forecasting be tied to the overall project schedule. Some equipment vendor contracts can be set up to pay fixed amounts on specified dates. This makes the cash flow forecasting somewhat easier.

Figure 2.5 is a generic version of a spending forecast form that can be used.

3.2   Overall Project Schedule

The overall project schedule is composed of the following three schedules. Milestone dates for critical events from each schedule will be used for approval documents. You will have to have thought through the three schedules to arrive at reasonable milestones.

3.2.1   Engineering Schedule

This is a schedule showing the duration of the engineering design. Whoever is doing the engineering design should supply this information. All vendors should provide schedules that reflect engineering, manufacturing, ship dates, and installation time. All this information should be compiled into one engineering schedule.

3.2.2   Procurement Schedule

This schedule shows when equipment has to be purchased to meet the project schedule. Vendors can supply equipment delivery dates from which the schedule must be backed up to determine when to start engineering on those items. Some of these dates will determine shutdown dates

PA # _____ DATE: _____			
Approval + 1 Month	$ _____	Approval + 13 Months	$ _____
Approval + 2 Months	$ _____	Approval + 14 Months	$ _____
Approval + 3 Months	$ _____	Approval + 15 Months	$ _____
Approval + 4 Months	$ _____	Approval + 16 Months	$ _____
Approval + 5 Months	$ _____	Approval + 17 Months	$ _____
Approval + 6 Months	$ _____	Approval + 18 Months	$ _____
Approval + 7 Months	$ _____	Approval + 19 Months	$ _____
Approval + 8 Months	$ _____	Approval + 20 Months	$ _____
Approval + 9 Months	$ _____	Approval + 21 Months	$ _____
Approval + 10 Months	$ _____	Approval + 22 Months	$ _____
Approval + 11 Months	$ _____	Approval + 23 Months	$ _____
Approval + 12 Months	$ _____	Approval + 24 Months	$ _____
SUB-TOTAL	$ _____	**SUB-TOTAL**	$ _____
		TOTAL	[_____]
Project Engineer _____ Date: _____			

Figure 2.5 Generic Spending Forecast

3.2.3   Construction Schedule

After the above two schedules have been developed, the construction schedule can be put together. It should reflect the equipment deliveries and when the engineering design is completed. If needed, talk to contractors to get accurate estimates on installation.

After looking at all of the above items, you will now have a fairly comprehensive cost estimate for your project. During the approval process you should have no difficulty in determining what is included in the estimate and you should be able to answer any questions quite readily. Once the project is approved you can convert the estimate into your project budget without too much rework. At the start of construction use the budget code numbers to prepare the progress payment forms for the contractors.

Chapter

3

ENGINEERING CONTROL AND DESIGN

Introduction

This chapter looks at how to set up a filing system for your engineering office and discusses issues related to engineered drawings. The areas discussed are:

- Engineering files
- Drawings in an engineering office
- External drawing issues
- Drawing revisions
- Drawing transmittals
- Document distribution
- Drawing certification

Engineering Files

The plant should have a central file system already in place. In order to cut down on the amount of paper on your desk and in your personal files, you should get in the habit of using the central file for filing as much material as possible. Generally, it doesn't make sense to copy pages of information just to keep in your personal files that will never be looked at. There are some items that you may want in your own files, but you should be selective. Usually, copying is a waste of time and money.

The file system should be easy to follow and understand. If you have information you don't know where to file, add a new file number that makes sense. Make sure you add the number to the master file list as well. (The master file list is the numerical list that relates the file number to the file name). The filing system should be fluid with the ability to add numbers as required. It is your responsibility to put the appropriate file number on a document. Leaving this up to a secretary or file clerk can result in information getting lost.

Request For Quotations (RFQ's), Purchase Orders (PO's), and Contract Documents (CD's) are usually filed separately from the rest of the files, i.e., they do not follow the standard file numbering system from the master file list. Following is a generic description of the numbering and filing system for RFQ's, PO's, and CD's.

As discussed above, the accounting group gives a number to all projects submitted for approval. This number should appear on all documents sent to the central filing system. Different companies use different numbering systems, but typically each part of the number assigned has a specific meaning within the company. A typical project number might be PA-S112. The breakdown is as follows:

- PA = Project Authorization

- S = a project in the Stores Area. (Alternatively, a Major Area Number might be used instead of a letter, e.g., PA - 57-112, where 57 is the number for the Stores Area. This depends on how the plant is set up.)

- 112 = the project number assigned by the accounting group meaning project #112 of a running series of projects in the Stores Area.

Within the filing system there should be a separate file drawer for PA – S112 where the above RFQ's, PO's and CD's can be filed. Documents to be filed in the general file will have PA – S112 on it plus a file number off the master file list (i.e. PA – S112 / 6.8 could be an internal memo regarding project PA – S112). Once the project is approved, the PA is dropped and a filing system for S112 is set up in the central file area. S112 becomes the project number.

The purchasing department will assign an RFQ number. When you give the RFQ to purchasing, they will assign a purchase order number to the RFQ. On large projects these purchase order numbers are determined well in advance and will show up on a purchasing schedule. These numbers will be the same from project to project, i.e. 1001 is structural steel, 1002 is civil construction, 1003 is mechanical installation, etc. The purchase order number/contract number has additional numbers (or letters) added to make it into a RFQ number. for example (using project S112):

- Purchase order number/contract number 1021 (assume this will be for painting) is going to be issued by purchasing. Before a purchase order / contract can be produced, one has to go out for bid using a RFQ.;

- The RFQ number is made up and may be of the form S112 - 100 - 1021; again, if your plant is set up with major areas, the RFQ number might become S112 - 100 - 57 - 1021 with 57 being the Stores Major Area. This way you can look at the RFQ number and know what area of the plant the RFQ applies to (you now have a RFQ for painting in the store's area).

- Upon contract award the 100 is dropped. The 100 is only used to signify that the document is out for bid;

- The purchase order number/contract number becomes S112 - 1021 or S112 - 57 – 1021 (you now have a purchase order number / contract number for painting in the stores area).

The additional numbers are added for control purposes. Do not issue PO numbers that are exactly the same as the RFQ numbers. The control numbers are added so that when discussing projects you know exactly what document you are talking about and what stage of the project it is in.

Prior to contract award all price inquiries, RFQ's, quotations, bid evaluations, data, and information from all vendors should be filed in the project authorization file. After contract award, all documents relating to the unsuccessful vendors should be transferred to a holding area in the files. Do not throw them out as you may have to refer to them. Filing will then be within the project by purchase order number if more than one purchase order is written. If only one purchase order is written, file by the project number.

Upon completion of the project the project files should be purged. Unsuccessful bidders information should be destroyed and successful bidder information transferred to the appropriate file according to the standard file system. Check with your accounting group to determine what information they have to retain for income tax purposes before destroying anything. The following are some examples of how various items are filed:

Examples:

PA S112/1.6.1   A letter in S112 before project approval filed in 1.6.1.

S112/00335/1.6.3   A fax under PO 00335 within Project S112 filed in 1.6.3

S112/00442/1.6.1   A letter under PO 00442 within Project S112 filed in 1.6.1.

Vendor Equipment
(Documents, Correspondence, & Drawings)

Vendor equipment information should be filed by purchase order within the project number and if possible double filed by the standard file system number.

Contracts

All contract information and documentation should be filed under the contract number within the project number.

Engineering Study File

A file should be kept of all completed engineering studies. Such studies would include borehole test reports and major engineering studies for productivity improvement, etc. These studies can be cataloged as follows:

	Denotes:	88	57	012
88	Engineering Study			
57	Major Area			
012	Serial Number			

An index is kept in the first file of this group of files. The title, consultant's company name, report number, author, and date are also recorded. The engineering secretary or other approved person assigns the index number.

Consultants File

A file of engineering consultants is usually maintained in alphabetical order. Each company resume is dated so that when superseding company resumes are received, the outdated resumes can be discarded. The following are some categories for consultant files:

Civil   Mechanical
Environmental   Structural
Electrical/Instrumentation   Other

Contractor File

A file of contractors is maintained in alphabetical order by category, not by company. Again each company resume is dated so when superseding resumes are received the outdated resume can be discarded. The following are some categories:

Civil	Piping
Electrical/Instrumentation	Structural
Mechanical	Other

Vendor File

A vendor brochure file is also kept in the engineering area. This information is filed by type of equipment and/or vendor service, not by vendor. You should date and put a file code on all literature deemed worthy of keeping for future reference. Following is a suggested list of vendor headings for an industrial plant:

• Agitators	• Heat Exchangers
• Air Compressors	• Hydraulics
• Air Conditioners	• Instrument Equipment
• Boilers	• Piping
• Cleaners	• Pneumatics
• Coatings	• Presses
• Conveyors	• Pumps
• Electrical Equipment	• Structural Components
• Evaporators	• Tanks
• Fans	• Vibrators

Plant Equipment Files

Documentation on all plant machinery should be kept in the plant equipment files located in the engineering area. The maintenance department also keeps a duplicate set. These files are indexed using the equipment number.

Where there are several pieces of duplicate equipment (e.g., Dryer Fans – there might be 20 or more of these) and each has a different equipment number within the same major area, only one file is kept. All documentation is kept in the equipment file with the lowest number.

General Files

The general file includes the filing of letters, faxes, and other such correspondence. This book does not cover the general file as there are too many different systems available and every plant office has one in place. The preceding file information is something that may not be in use and can be put in place quite easily.

Drawings

Engineered drawings are the major source of project information, so the plant-engineered and vendor-engineered drawings should be produced according to the plant standards and procedures. If these standards and procedures are followed, the process of extracting information from the drawings becomes fairly easy. Make sure that everyone involved in drawing production, including those outside the plant, follows your procedures.

The following are some important drawing issues you should be aware of:

a) During the engineering phase, make sure you have a drawing stick file and procedure in place for revising drawings. The stick file is a hanging file of the latest issue of drawings. Revisions should be marked, in red, on the stick file drawings and the person making the revision should initial and date the revision. Do not use any color other than red and do not let people mark your drawings with anything but a pencil since only pencil can be erased. You will find that people love to look at drawings and discuss issues. Invariably someone has a pen and will try to mark up your drawing with it. Don't let them, give them a pencil instead.

The drawing revisions are allowed to collect on the drawings, and once there are enough to make an official drawing revision worthwhile, the original drawings are updated. Ignore all revisions that are not in red and not signed or dated. Your engineering office personnel should follow this procedure also; otherwise, you will loose control of the drawing production process.

When reviewing or checking drawings use a yellow pencil to cross out the items you have looked at. If you don't cross off what you have looked at, you will keep going over the same items and won't know when you have looked at everything.

b)   When it comes to checking engineered drawings, you have to be very careful. Typically, drawings are prepared and then given to a "checker." This is an experienced person who goes through the drawing checking that the dimensions are correct, the drawing has been produced to the applicable codes and standards, the border information is correct, etc. Part of the title and issue block is a section for the initials of the persons who designed, checked, and approved the drawing. If you look at some of the drawings you currently have in your office, you will notice some on which the designer is the same person who checked the drawing. This should not be allowed to happen. The checker should always be someone other than the designer.

When preparing a drawing you become absorbed in the drawing and do not notice your own mistakes. You may repeat the same errors over and over without realizing it. For this reason, you need a fresh pair of eyes to look at the drawing, check the dimensions, check calculations, check information, etc. This is what the checker does for you. Similarly, if you are looking at a drawing another person is working on, point out any errors you notice without being critical. Even when producing only one or two drawings, checking by others should take place. However, keep in mind that the drawing should be checked, not redesigned. There could be a tendency on the part of the checker to redesign what has already been done. This is a waste of time and effort and should be avoided.

c)   You should send copies of your drawing standards and procedures to the vendors producing drawings for you. Do not assume that they understand your drawing standards or are using a system compatible to yours; otherwise you could end up with drawings that:

- are not compatible with your CADD computer system;
- are on odd size paper and will be either too large or too small for your stick filing system;
- have the North arrow pointing in the wrong direction on the sheet making it difficult to read the drawings;
- have revisions that are hard to find.

d)   There are standard drawing sizes for different types of drawings. Instrument loop drawings may be generated on 11 x 17" sheets while mechanical layout drawings are generated on 24 x 36". Your vendors should try to keep to the sizes you request. No matter how hard you try, you will still occasionally receive metric size sheets. It depends on what country the drawings are produced in.

Some drawings for equipment cannot be made to the plant standard and will come on very large sheets, which you will have to live with. This is peculiar to the machine vendors.

e)   You have watch out for the direction of the North arrow on a drawing. Standard convention is for it to point to either the top of the drawing or to the left side of the drawing. You should insist that all drawings be prepared the same way. You need consistent orientation in your drawings; otherwise, it can be a problem in the design office and in the field trying to understand the drawings. You could run into rework problems, and if concrete has been poured, you could have a schedule delay. It is up to you to make sure the drawings are correct.

Case History 12

Before pouring a major equipment foundation, a surveyor was called in to double check the location of anchor bolts that had been set by the contractor according to the vendor's drawing. The surveyor was familiar with the site and knew where site North was. He had trouble orienting himself with the drawing and upon further investigation it was discovered that the North arrow on the vendor's drawing was actually pointing South and the bolts had been set backwards. Fortunately no concrete had been poured and the cost to the vendor of redoing the bolts was not too great. In this case the vendor had not followed drawing procedures and the consultants did not check the drawing well enough.

f)   Dimensions on drawings can be a problem if you require them in metric. Europeans will give you a metric dimension that is a real number since they work in metric. Dimensions from US firms will be soft conversions. A soft conversion is taking an imperial unit and converting it to metric. Soft conversions will give numbers like 521 mm or something similar. (1-5/16" = 1.3125" = 33.3375 mm or 33.3 mm.) You will not be able to get the correct imperial unit when converting back. (33.3 mm = 1.311") This may look okay on a drawing but in the field you will not be able to measure 1 mm. Rather than making soft conversions, it may be more prudent to build in imperial units to make sure everything is right. If you are doing a straight metric project, you should insist that all dimensions be taken to the closest 5 mm, e.g., 521 mm would be 520 mm and 643 mm would be 645 mm, etc.

Case History 13

On a plant being built in Canada (in metric) we had a US firm supplying a complete equipment package. The equipment drawings were in imperial units with a soft conversion to metric. From a construction point of view, the converted numbers did not make sense. In order to make the drawings comply and to be able to manufacture and install the equipment, we had the drawings dimensioned in imperial and soft converted to metric. All the equipment was manufactured and installed in imperial. There were several survey tie-in points for the equipment to the building, which we made in hard metric numbers.

g) If your plant is not on CADD, you should supply your vendors with blank mylars bearing your plant's title block and logo. If your plant is on CADD, the title block and logo can be provided on disc to any vendor that requires it.

h) For CADD produced drawings, your plant CADD drawing procedure should be given to the vendor. This will ensure that you get computer compatible drawings, i.e., that the levels are the same color, line thickness are the same, the same CADD version is used, etc. If your standards are not being followed, your CADD operator will have to convert the electronic drawings to your standard before printing to have consistency in the drawings.

i) All drawings produced that are going outside of your engineering office should have a drawing number or a sketch number on them. These series of numbers are usually kept in the plant engineering office. Assigning numbers to drawings is usually a draftsperson's responsibility. The numbers are maintained in a log book that records the number, drawing title, date issued, and the name of the person taking the number out. The numbers are always assigned when a drawing is started.

If you are making a sketch as part of a bid package, the sketch number should include the project number plus whatever additional numbering convention you want to use. Make the number unique so there is no mix up with other drawings. As an example:

Sketch Number: 1021-mhs sk-1
    1021 = Project Number
    mhs = initials of person who created the sketch
    sk-1 = Sketch #1

j)   For small jobs being done by a vendor, a block of numbers can be given to the vendor to use. This keeps the drawing numbers in sequence and keeps all the individual vendor drawings in groups. If you don't do this, you will end up with in-house engineering drawing numbers, or even other vendors' drawings, mixed in with a vendor numbering system. This makes for a messy filing system and it makes finding details a lot more difficult. When giving out blocks of numbers, leave some numbers between groups in case a vendor requires more numbers.

On large jobs and on equipment drawings the vendor drawing numbering systems can be used and then converted to your numbering system. This is done by modifying your title block to allow the inclusion of the vendor's drawing number over yours. The drawings would then be tracked using your numbering system with a cross-reference to the vendor's number. With electronic drawings it is just as easy to place the vendor drawing onto your drawing sheet without changing the title block.

k)   Ensure that all drawings are cross referenced on both the layout drawing and the detail drawing. On the right side of the drawing sheet is a column titled "Reference Drawings." For example, the layout drawing should reference the drawing the details are on and the detail drawing should reference the layout drawing that the details relate to. By cross referencing, field personnel can find the required drawings and details quite readily. If someone sends you drawings that are not cross referenced, go back to whoever produced the drawing and get the situation corrected immediately.

l)   All plant drawings have a numbering system derived from the consultant who originally built the plant. They are, in one form or another, derived as follows:

•   The plant will be broken down into areas and each of these areas will have a number, e.g., the boiler area may be Area 25.

- If you look at drawings within the area, you will notice they are numbered according to what department produced the drawing, e.g., Structural #100, Mechanical #200, Piping #300, Electrical #500, etc.
- There may be a further breakdown into material handled (water, oil, air, steam, etc.) or sub-areas which will take a block of sequential numbers.
- The drawings will be numbered sequentially within each sub-area. If a block of numbers from 30 to 60 was taken out for steam lines, then drawing number 25-351 could be a steam piping drawing in the boiler area.
- It is important that you understand how the original plant drawing numbering system works. When you have a project consisting of several hundreds of drawings, it makes finding information a lot easier.

m) For in-house drawings you should establish the drawing matchline locations for drawings to be used by all disciplines. (Matchlines are used when an area has to be covered using several layout drawings. The layout information is drawn up to the drawing matchline on one drawing and starts at the matchline on the next drawing. When you put the matchlines together, the drawing information continues across the drawings unbroken.) Using consistent matchlines ensures that all drawings reflect the same process area and orientation. For drawings done by outside consultants they should conform to the consultants' own procedures. You should inquire if they have one. Make sure equipment suppliers follow a similar procedure.

At the matchline, the next drawing should be referenced with words to the effect **"For continuation, see Drawing _____."**

n) Key plans should always be made and located on the bottom right hand corner of all drawings. This reference shows what area is covered on the drawing. This makes searches easier as you look through the key plans until you find the appropriate area highlighted on the key plan. Some drawings are produced with key plans in opposite corners, which can be handy at times.

o) Never reuse canceled drawing numbers for any reason. If the number is canceled, mark it as such in the book and cross it off so it cannot be used again. You do not want drawings with the same number being issued. This applies to revisions as well. If you come across this problem, correct it immediately.

p) Once a drawing has been issued, do not change drawing numbers unless you absolutely have to. This makes it too confusing for everyone involved.

Drawing control is very important in any design office. It does not take much to get into trouble with mixed up drawing numbers and revision errors. You should know and understand the plant drawing procedure and follow it. If the procedure is wrong, change it to something workable. Do not take shortcuts with drawing control.

External Drawing Issue

Any drawing you issue to others outside the plant organization (e.g., to consultants, vendors, contractors, or others) for any purpose is classified as an "external drawing issue." It is your responsibility to issue all drawings for approval, fabrication, Certified For Construction (CFC), tendering, and contract. This responsibility can be delegated, but you should sign off (or approve) the drawings, and you should know what has gone out and to whom. When issuing drawings make sure a transmittal is used and that it indicates the drawing number, issue number or revision number, and the drawing title. (Write out the drawing title; do not use abbreviations.) This is your record of what was sent out, so do it right.

If the drawing has changed in any way it should have a revision number or issue number. Even the slightest change should be treated as another revision and assigned the next revision or issue number. Do not try to sneak a minor spelling or other change through without a revision. Typically you would save change items and then make a revision when you have several of them, provided the changes are not imperative.

If a change is small but important, make a sketch of the change and issue the sketch. Revise your stick file drawing and when there are enough changes the drawing can be revised to catch up with all the other changes. Some companies use a system of "Design Change Request" for drawings that have been issued "Certified For Construction." With this system, a standard form is used on which the details of the change are drawn and the forms are issued using a sequential numbering system that refers back to the original drawing. These change requests are treated as drawing revisions by the construction site and the changes can be implemented immediately. The design office collects the change requests and after approximately 10 are collected for a drawing, that drawing is revised and issued.

One copy of all issued drawings should be folded and kept in a file so you have a copy of what was sent out. Do not depend on any other method. Get a file copy for yourself at the same time the other prints are made. You should also maintain a master stick file that has only the current working drawing on it. A copy of all drawings issued for tender and construction should be filed. If you issue a large tender package you, can keep one set of tender drawings on a stick file for easy reference. All drawings being issued should be date stamped after printing with the current date. The drawing on the master stick file should be the latest issued drawing. From your drawing files you should be able to build a legal case from the first drawing to the last drawing.

Even being this careful you will always run into the odd glitch. If you run into problems, look at the situation, see if there is a procedure on how to handle it, and make the changes necessary to get back to what the procedure describes. If you have to make another issue of the drawings, then so be it.

Drawing Revisions

Revisions are changes and/or additions to "technical" information on
drawings. Each plant and consultant will have a specific procedure
for drawing revisions. One method is to use issue letters until the
drawing is "Certified For Construction" (CFC; or AFC–"Approved For
Construction"). Once the drawing has been issued CFC, then revision
numbers are used. The first issue of the CFC drawing is Revision 1
not 0. There is no Revision 0. You will go from the last issue letter to
Revision 1.

Once a drawing has been issued you must make sure that any
revision descriptions noted in the revision column are correct, dated
and signed, and the correct revision number is shown in the revision
block and in the box next to the drawing number. Make sure you
check all this information before prints are made. A mistake with
revision numbers or other drawing issue information can create
problems after the drawing has been issued. So get it right before it
gets printed!

There is no such thing as "General Revision." All revisions should be
clearly indicated on the drawing and described in the appropriate
space. When a revised drawing is received on site or at the
manufacturer, the revisions are noted and the work carries on. When
you issue a drawing as a General Revision, someone has to spend the
time to check every detail to see what has changed. This is poor
engineering and laziness on the part of the engineer/draftsman. You
can get backcharged by contractors if they have to spend too much
time looking for revisions on "General Revision" drawings.

Method

A typical drawing revision procedure is as follows:

- The change will be described in the revision column with a
  clear and concise description of the revision.

- The revision will be identified correctly on the drawing (see "Identification of Revisions" below).

- A sequential numeric code is used. The first revision will be 1, the second 2, and so on.

- At the time of issue, you will determine that the drawing revision is properly closed off, i.e., the revisions have been checked and approved.

It is important that drawings be checked and approved and the initials put in the respective boxes on the drawings. You should not issue drawings "CFC" that have not been checked or approved.

Identification of Revisions

To cut down on the time and cost of locating drawing revisions, revisions are usually indicated with flags and by clouding (see Figure 3.1). This may seem like a trivial matter but on large piping drawings it becomes very important as the drawings can be cluttered and revisions difficult to find.

A typical revision procedure follows:

- Each change or addition should be clearly clouded and identified with a triangular flag containing the current revision number. The clouds are drawn on the back of the tracing with a soft pencil. Lines circling the previous revision should be removed. This is done with CADD drawings as well. On CADD drawings the revision is clouded on the computer file which shows up on the drawing.

- The flags are placed next to the spot where the change was made. To clearly indicate its location, additional flags are placed along either the right hand or lower outside border of the drawing. Aim the flag by pointing one corner of the triangle toward the change.

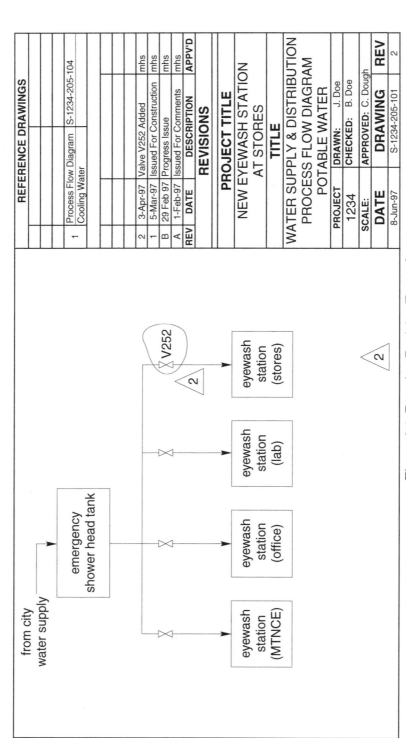

Figure 3.1 Drawing Revision Example.

Drawing Transmittals

It is important to keep a record of drawings and documents transmitted to others outside your plant. This record keeping is done with transmittals. The transmittal is a form or cover letter that describes the drawing or document being sent. Use an "Outgoing Transmittal" file and an "Incoming Transmittal" file for easy retrieval of transmittals. Do not throw out a transmittal; you never know when you will need it. Transmittals should be used for everything sent out of the office including accounting information, invoices, etc. All items should be clearly listed with reference numbers. When you are sending material to and from a construction site you will normally use a courier. These courier packages can get lost so it is imperative you know what you are sending out.

For drawing issues external to the plant, the plant procedure should be followed. Again a record of what was sent and to whom it was sent should be kept. Always send out drawings with a reason stated on the transmittal. There are usually check-off boxes for this purpose, or you can write instructions on the transmittal. Follow up as required on drawings that are to be returned.

If drawings are being sent by courier, the courier receipts should be kept for tracking purposes. Put as much information as possible on the shipping forms so that retrieval of the correct form is easier. Identify it so you can look back at the slip in a week and be able to tell what was in the package. Be assured that on any given project, packages will go astray and you will be looking for what was sent out and when. This collection of information is just another method of protecting yourself by having the necessary information at hand and knowing you have done your end of the job properly.

You are responsible for the correctness of all drawings and other material sent out of the office, and you should be aware of what the drawings are for and to whom they are going. Do not depend on others for this.

Document Distribution

Work off a document distribution list when issuing materials. This is a list you make at the beginning of the project indicating who gets what drawings and other information (see below). It helps everyone in the plant know, on a continuing basis, who is supposed to get what drawings and information and how many copies.

Figure 3.2 is a generic document distribution form that can be modified to suit your situation. The easiest way to determine who gets what information is to send a blank form to everyone on the list and ask them to fill in what they want. Give them a deadline for responding. Review the responses and decide for yourself what information to provide to each person.

Because you do not want to waste time and money making up documents for someone who wants information because it makes them feel important, there is one way to find out if the recipient really uses is the document or not. Stop sending it and see what happens. If you get no response, you know they were not using it. Now you can cut that person off the list.

Give the final document to your secretary to distribute. When you get a document to be distributed, classify it and give it to the secretary marked "For "Distribution." He or she then uses the document distribution list you've developed to distribute it. This saves you from writing out all the names of the people to whom it should be sent. Some plants have a stamp made with all the names on it. You stamp the document and then simply check off the names of the people you want the document to go to.

Description	Plant Manager	Project Manager	Project Engineer	Mechanical Design	Civil Design	Electrical Design	I & C Design	Environmental	Purchasing	Stores	Project File
Flowsheets		X	P	P	P	P	P	P			P
Layouts		X	P	P	P	P	P	P			P
Vendor Drawings		X	P	P	P	P	P	P			P
Tender Documents		X	C	C	C	C	C	C	C		C
Quotes		X	C					C	C		C
Project Scope Documents	C	X	C	C	C	C	C	C	C		C
Project Status Reports	C	C	C	C	C	C	C	C	C		C
Purchase Orders		X	C						C	C	C
Internal Correspondence		X	C								C
Transmittal In		X	C								C
Transmittal Out		X	C								C
Government Correspondence	C	X	C					C			C

Legend: C - Copy   P - Print   X - Circulation

Figure 3.2 Document Distribution Form

Drawing Certification

The plant should have a procedure set up for issuing CADD drawings. Each drawing revision is a new print and the print does not reflect the signatures of the previous issue (unlike tracings). The CADD operator will have to put the correct initials in the appropriate approval boxes in the electronic file. A lot of drawings will have an engineer's stamp on them. The first issue of a CFC drawing will have an engineer's stamp on it, and subsequent issues of the same drawing will have a reference to the stamp. This is acceptable as long as the stamping engineer checks and approves the revision. For drawings issued outside the plant that need a stamp, such as government agencies, each print will have to be stamped and signed. Any further issues of these drawings will also have to be stamped and signed. This is called wet stamping.

Now that you have the drawings made and approved it is now time to assemble the drawings, make a request for quotations, and purchase the equipment or service.

PROCUREMENT

Request for Quotation - General

Before anything can be purchased for the plant, you will have to get a price quote on the item or service from a vendor or contractor. The standard method of getting this price quote is to use a request for quotation (RFQ). As you will see later, the documentation you send out will have to define clearly what you require. The information is sent to the required number of vendors and if the process is followed correctly, all vendors receive the same information and are bidding on the same thing. When the required information comes back, make sure you are comparing "apples to apples."

Case History 14

I was working for Central Engineering, and got involved in a project about half way through it. The plant had begun a major machine rebuild and were now in well over their heads. They had purchased a used plant in the UK and had entered into a contract with a demolition company. The RFQ scope was poorly written and the contractor had them over a barrel. At every major step in the removal of equipment the contractor demanded more money before he would proceed to the next phase. He destroyed part of the building getting some of the equipment out (he dropped a forklift through the roof to make an opening) and he destroyed some of the equipment to get it out. (He actually cut the trucks of a bridge crane in half to get them out, then crated them and put them on the boat.) The project was overrunning so bad that the scope was cut to what was affordable. I sat in on one meeting where the plant project manager had to tell a senior VP that the overrun had increased by $750,000.00 in one week! The equipment eventually came over on a boat (it was a pile of junk), was stored in various locations around the plant, and was never used. This was a case of not defining the project scope and signing a contract based on this poorly written scope.

Every year accountants audit your companies' books. They are making sure the business is run according to accepted business practices and operating within the law. By issuing the RFQs and using the bidding process, you are proving to them that the plant is following these standard accepted business practices. The RFQ is just one item of good business practice you should be following. You are also showing the company shareholders that the plant is getting the best value for the dollars spent.

The bidding process keeps you honest; that is, no one can accuse you of just giving work to your friends or showing favoritism to others. You should only give work to your friends if they have followed the bidding process and you can honestly justify using them over others. This aspect of the bidding process is more important if your plant is in a small town and the plant is the major employer. Keep your business relationships above board. In small communities, some business people do not understand the bidding processes and assume favoritism for others when they loose out on work.

You should keep the bidders list to a minimum. Usually three bids are required. Sometimes you will only find one or two qualified bidders and other times you will find too many qualified bidders. It makes no sense to get lots of bidders as it only makes more work for you and there may be no cost advantage. Deal only with bidders you are happy with and want to deal with. Government projects let anyone bid. There are no restrictions on the number of bidders, and the bidders have to pay for the bid documents. Privately owned businesses do not operate on the same principle and you can limit the number of bids and choose who you want on the bid list.

It costs money to prepare a bid and for large projects it can cost the bidder upwards of a $100,000.00. It is not good business practice, nor fair, to ask a bidder to quote if you do not intend to use him. In the long run you can sour business relationships for a long time by doing this. With the increase in the number of EPC contracts being used, some companies are paying the bidders to prepare a quote. This way, the owner can get qualified bidders to bid, who may otherwise not bid because of the cost involved in putting a bid together.

Request for Quotations - Types

You will have to work with basically two types of RFQs. They are:

Type 1:

Those that are for equipment only with no manpower being supplied on site (other than start-up people), e.g., pumps, screens, heat exchangers, complete process package of several pieces of equipment, etc.

Type 2:

Those that are "supply and erect" or "erect only." Supply and erect involves a contractor supplying equipment, material, and manpower to install the equipment. Erection only involves the contractor supplying the manpower only to install equipment and material supplied by others, e.g., piping system, boiler package, etc. These types will be discussed in greater detail under "Construction Management."

RFQs should be sent out only by the purchasing department or by you under the direction of the purchasing department. This is important as the required purchasing forms and terms & conditions (T&Cs) have to go out with the documents.

The purchasing group will help ensure that any information that is going to affect the contractor's price is in the RFQ documents. The purchasing group will have a better understanding of the bonding and insurance requirements. If these items are required, this type of information has to be given to bidders before contract award. If you provide this information after contract award, you could be charged for an addition to the scope or an "extra" to the contract. An extra is any additional work that is over and above the signed contract scope. If you agree that the work is an extra, the payment of the extra is handled with a Field Work Order (FWO).

To give bidders sufficient information to respond with a meaningful price and to make sure you have control of the project, the RFQs should contain the documentation listed below.

Type 1 RFQ

This RFQ should contain as a minimum the following documents:

a) Purchasing's Request for Quotation form (if available) including plant terms and conditions
b) A Scope of Work, schedule or milestone dates, any specifications, drawings, and standards
c) Plant conditions and standard component list
d) Vendor data requirements
e) Vendor information requirements
f) Equipment specifications
g) Material specification
h) Any other information relevant to the equipment/project.

Type 2 RFQ

This RFQ should contain as a minimum the following documents:

a) Purchasing's Request For Quotation form including plant terms and conditions
b) An RFQ form or a tender form that includes a Scope of Work, general conditions, special conditions, schedule or milestone dates, drawings, and standards
c) Material specifications
d) Equipment specifications
e) Any other information relevant to the project.

For both Type 1 and Type 2 RFQs the following documentation will have to be generated:

An RFQ Title and Number

As discussed previously, many RFQ numbers eventually become the purchase order numbers. These numbers can refer to different plant areas and type of contract. It is important that a number and title be issued for a RFQ, as the number will be referred to throughout the life of the project. Titles help others who do not know how the numbering system works find and organize the project documentation. The title should be clear as to what the RFQ is for.

List of Bidders

You should put this together with input from the purchasing group as required. Keep the following points in mind when making up the lists:

If you don't want to do business with a specific vendor/contractor, don't put him on the bid list. It always turns out that the one you don't want to deal with has the lowest price and you then have to justify to management why you don't want to deal with him. Say no to the vendor/contractor up front.

Be leery of contractors recommended by upper management or those that are friends of management. If they are awarded the contract, you will have trouble controlling the job, as they will always try to bypass you and deal directly with your superior.

Be aware that sometimes management does not want certain vendors/contractors or people on site. Always check with your supervisor about who to put on (or keep off) the bid list.

If in doubt about a vendor/contractor check their references and don't be afraid to leave them off the bid list. The aggravation of dealing with a bad vendor/contractor is not what you want to do nor have the time for.

If the job is large it may be advisable to do a credit check on the vendors/contractors or have them fill out a qualification form. Keep in mind that you cannot find out if the vendor/contractor owes back taxes to the government. You can check the contractor's references about the garnishment of payments for back taxes or ask the contractor directly about back taxes. He may not tell you but at least you tried. The issue of back taxes is not a common occurrence, but you may be involved with the issue at least once in your career.

If the contractor owes back taxes and the government finds out he is working, his payments will probably be garnisheed. As you can imagine, this is one headache you want to avoid at all costs. Unfortunately you can't do much about it until it happens. When it does happen you will have to get legal advice and will more than likely have to make arrangements to pay his subcontractors directly and bypass the contractor. What is left over from the subcontractors will go to the government. At least this way you can keep the project going.

Equipment Specification

You will have to prepare the equipment specification. The plant may have standard forms available, which should be used or followed. The form is important as it describes what you want and should address the following items:

a)   Scope of Work;

b)   What is included in the work?

c)   What is not included in the work?

d) Design criteria and what the equipment will be used for. It is important that the supplier know what the end use is so your quote is based on a suitable item. Your specifications may be wrong and this is a chance to have an outsider more familiar with the products correct the specification. If you get the wrong item and have problems after installation, at least you have some recourse, as the supplier knew exactly what you were going to use the item for and he should have supplied something suitable.

e) Material and component specifications should be as complete as possible with what you know at the time.

f) Equipment tagging. This is a standard size stainless steel numberplate that is attached to all equipment in the plant. The plant should have a description of a standard equipment tag on file. Equipment tagging also describes how all the shipping crates or loose items should be marked for easy identification by the receiver.

g) Proposal requirements. State how many copies of the proposal do you want so you don't have to do the copying. For large quotations it could cost you to get additional copies from the vendor.

h) Performance guarantee. If the vendor does not have a guarantee you agree with, develop one that everyone agrees on. Do not sign a contract or issue a purchase order unless you have a performance guarantee that everyone agrees on.

i) Warranty. If you want other than the standard warranty requirements, describe specifically what you want.

j) Acceptance. How will everyone know when the piece of equipment is performing satisfactorily and the supplier's warranty starts?

Before developing an equipment specification from scratch, check for past specifications in the files or library. There may be some specifications available from the original plant design and construction you can emulate.

Scope of Work

In preparing the Scope of Work, lead off with a general statement of what the work/equipment is and what the outcome will be. A more detailed description will be included in the following scope items:

a) **Included in the Work** - This is a detailed description of what is involved in the work or equipment supplied. It is important that this document is accurate, clear, concise, and covers all items required. If in doubt about any item, make a decision and include something and change it later if need be. Ambiguity will only cause problems later. Watch your grammar, use simple sentences, do not use abbreviations unless you state clearly what they mean, and make sure there is only one meaning to each sentence. Use point form or bullets if you have to.

b) **Not Included in the Work** - This is a description of work that will be done by others—either other contractors or plant crews.

Drawings and Specifications

The RFQ or tender package should have a list of all drawings and specifications being sent out. This should include the revision number and/or issue number. It is important you know what has been sent and to whom. It is dangerous to send different packages to different contractors/vendors. Send the same package to everyone. Keep a complete set of everything sent in your files until the end of the job. At contract award you will have to sign off on the contract drawings so make sure you have the correct ones. Do not leave it up to the contractor/vendor to have the correct drawings and specifications.

Schedule

With any RFQ or tender package sent out, there should be a milestone schedule included. This is especially true if time is of the essence as tight schedules could affect the price. That is, you will pay a premium for a short schedule.

Standard Documents

Standard documents the plant should have on file are the plant conditions and standard component list, vendor data requirements, and vendor information requirements. These are required in the RFQ and should be sent to vendors for equipment supply and should be given to the vendor before contract or purchase award. The cost of these items should be included in the bid. The plant may have these documents under other names but they are as follows:

a) **Plant Conditions and Standard Components List** - This lists the plant systems and general plant information the vendor will require in the design of his equipment. It will also list all the plant components and the preferred manufacturer preferred by the plant. This ensures that the plant gets components it is comfortable with and that are currently being used. Rest assured you are bound to get something other than what is listed, but if you have sent this document you can tell the vendor to take back the offending item.

b) **Vendor Data Requirements** - this document outlines what the plant expects the Vendor to do and provide. This information should be included in the Vendors proposal. Some of these items will cost you money if you give out this document after contract award.

c) **Vendor Information Requirements** - The information requested should be returned with the vendor's bid. This information is of interest to you and will be used in the bid evaluation. The answers to some of these questions could possibly disqualify the vendor.

Following are examples of the Plant Conditions and Standard Component List, Vendor Data Requirements, and Vendor Information Requirements. They can be modified to your particular case.

Plant Conditions and Standard Component List

This document is to be sent out with all request for quotations and tender documents that require a vendor or contractor to supply components as part of his equipment supply. It is up to the RFQ originator to make sure the vendor has this document before submitting a bid. Using this document ensures consistency in components throughout the plant. This prevents the disappointment of equipment arriving on site with components you neither like nor stock in your stores.

Be aware that when you buy a complete package from a vendor, he may be including components he gets a price break on or using a certain manufacturer because the components are reliable. The vendor should advise the plant if he does not intend to use the preferred component and explain what he intends to supply. This, of course, is always subject to the plant's approval. Be careful not to make any warrantees or process guarantees null and void by specifying a component the vendor is not happy with. Don't use the wording "this widget or equal" in your documentation. What you consider equal and what the vendor considers equal may be two different things. There is no "equal;" the terminology to use is "approved alternative," and the approved alternative must be approved by you in writing.

Plant Conditions and Services

The following wording should be included in the standard component list to give the bidders some guidance on how the information should be used. The author's guidelines are enclosed in brackets and shaded like this { }

The Plant Conditions and Services section describes the conditions and services the plant uses and provides. The information is freely offered; however, where the bidder's guarantee is dependent on the accuracy of this information it is clearly understood that the bidder is solely responsible for independently investigating and satisfying the bidder's requirements for such basic information.

The bidder shall take note of the plant's preferences where specifically related to the materials and equipment offered and the bidder shall confirm acceptance of these requirements or clearly state otherwise.

Guideline 1

{At any vendor meetings, make sure the vendor agrees to use the components the plant wants; otherwise he should tell you what he does not agree with.}

Where a specific instruction is given in the Requirements and Specification section, it shall take precedence over the following conditions, standards, and codes.

Guideline 2

{If you require a specific component, standard, or code for a one-off project and it is different from what is listed in the standard component list, state in the specification section of the RFQ what you require rather than revising the list.}

(Items in **bold and italics** should be changed to match your plant conditions. You can delete the metric or imperial, depending on what units your country works in.)

Plant ground floor nominal elevation above sea level is approximately **000.0** m or **000.0** ft.

Guideline 3

{Because plant floors slope, the ground floor elevation is taken at either a doorsill, the top of a floor trench, the top of manhole, or if you have them, at survey monuments encased in the floor. Consult your plant drawings for the correct location. The elevation can also be a mark you have run in. Examples would be the baseplate of a certain column, a fixed spot on a floor, location on a piece of equipment, etc.}

Guideline 4

{High and Low Design Temperatures: High and low temperature are needed for design of components to heat the buildings, such as makeup air units. If you are designing for the north, you have to have a low design temperature for steel to avoid brittle failure.}

Guideline 5

{Water Header Temperature Range: This range is required if you need a certain liquid temperature for a process requirement. During the winter the water temperature can be a few degrees above freezing and in the summer the temperature can be in the high 70s. You have to know the temperatures as you may have to add heat to the water or a chiller for cooling.}

Guideline 6

{Plant Water Pressure at Ground Floor Elevation: you need this information if you have to do pump calculations, or need water at a higher elevation. If possible this should be checked every few years.}

Outdoor Temperatures

Summer extreme high:	*00*°C Dry Bulb	*00*°C Wet Bulb
design:	*00*°C Dry Bulb	
Winter extreme low:	-*00*°C	
design:	*00*°C	
Water header temperature range:	*0*° - *00*°C	
Temperature range of water supplied to plant:	Summer: *00*°C	Winter: *00*°C
Plant water pressure at ground floor elevation:	*000* kPa (g)	*minimum, at design flow*
Natural gas distribution pressure:	*000* kPa (g)	*inside buildings*

Process Steam Pressure and Temperatures

Low Pressure Steam	Nominal conditions at steam plant:	*000* kPa (g) at *000*°C.
	Operating range:	*000* - *000* kPa (g) and *000* - *000*°C.
	Design (maximum allowable) conditions:	*000* kPa (g) at *000*°C.
Medium Pressure Steam	Nominal condition at steam plant:	*0000* kPa (g) at *000*°C.
	Operating range:	*0000* - *0000* kPa (g) and *000* - *000*°C.
	Design (maximum allowable) conditions:	*0000* kPa (g) at *000*°C.
Plant & Instrument Air Pressure:	*000* kPa (g) (Range: *000* - *000* kPa (g).	

Guideline 7

{Natural Gas Distribution Pressure: Large plants will have a high-pressure and a low-pressure line. They will have their own pressure reducing station on site. If installing any gas appliances, such as gas heaters, the designers have to know the pressure they are working with.}

Guideline 8

{Low Steam Pressure: Low-pressure steam is typically up to 300 psi. The steam pressure will have an operating range and all systems have to operate within this range. The design conditions are what all systems have to be designed to. If you install a component that is less than the design conditions, it will have to be protected from the design conditions by safety relief valves.}

Guideline 9

{Medium Pressure Steam: Medium pressure steam is typically 300 to 600 psi. It could be saturated steam. The same comments as above apply.}

Electric Power for Motors	
200 hp and less (including fractional hp):	*600 volts, 3 phase, 60 Hz (nominal)*
250 hp and up:	*2400 volts, 3 phase, 60 Hz (nominal)*

Guideline 10

{Motors: The break at 200 hp or 250 hp is an individual mill requirement based on incoming power. All motors should be 600 V or 440 V (U.S.). Do not use 120 V unless there is no other option. The reason for this is that 120 V motors can be used around a house and are therefore readily stolen.}

Electric power for control circuits: *120 volts* from individual controlled transformers (nominal).

Structural

Structural design shall conform to the *National Building Code, or IBC 2000, latest edition.*

The earthquake seismic data is $Z_a = 0$ $Z_v = 1$ *zonal velocity ratio 0.05.*

Wind velocity pressure *"g"* = *0.37 kPa for the probability of being exceeded once in 30 years.*

Ground snow load = *2.2 kPa. (for additional modification factors see National Building Code)*

Annual precipitation = *000 mm*
Greatest 24 hr. rainfall = *00 mm*
Design 15 min. rainfall = *00 mm*

Guideline 11

{National Building Code or the International Building Code (IBC): IBC 2000 has recently been accepted by some jurisdictions, so you may want to check your area. Use the words "latest edition" to signify the edition in force. This way you do not have to keep revising the document to reflect the most recent edition. It takes several years for the new building code to be reviewed and adopted by local governments, and you can only use the edition that has been approved by the authorities. Check your plant's original design specifications to see what was used.}

Guideline 12

{Earthquake Seismic Data: You should determine this value, not your vendor. These values will sometimes change with the Building Code, so you may want to review this number every time the Building Code changes. If a vendor does not understand the seismic data and uses the wrong factors, it can have a big influence on any structures you have designed. Check your plant's original design specifications to see what was used.}

Guideline 13

{Wind Velocity Pressure: This is used in the design of the building structural steel. For tall structures the choice becomes how much movement to allow at the top of the building and how much to spend on steel to limit the sway. A 300-ft. tall building with a movement of 1-1/2" at the top is not uncommon. Check your plant's original design specifications to see what was used.}

Guideline 14

{Snow Load and Annual Precipitation: Snow load is used to determine the building's roof and wall loading. The annual precipitation is used to determine the drainage patterns required for the roof and the size of the roof drain pipes. It is also used to determine runoff, ditch sizes, storm drainage patterns, manhole sizes, and culvert sizes. Check your plant's original design specifications to see what was used.}

Mechanical

Hub fits on standard machinery with similar shaft tolerances shall be ANSI Standard B4.1 interference fits.

Guideline 15

{You can find the ANSI B4.1 standard in most mechanical engineering handbooks. It is useful to read the standard and see what it entails. Other ANSI standards are referred to in this specification, so it would be handy to have them available as references.}

Couplings:
- Couplings shall be fitted flush with the end of the shaft.
- Standard couplings on drives 3600 rpm and higher: ***Gear type coupling.***

- Standard couplings on drives below 3600 rpm:
- *Elastomeric type coupling c/w QD bushings ≤ 100 hp*
- *Brand X grid type couplings >100 hp*

Spacer Couplings:

- Drives below 3600 rpm.
- *Elastomeric type drop out (provide on all horizontal pumps)*

Guideline 16

{If the coupling face and the end of the shaft are not flush, you will have trouble aligning the coupling. The spacer coupling with a drop out is required on all horizontal pumps as it will give you room to maneuver the motor or pump while on the base. The QD bushings mentioned allow for the quick disconnect of coupling from the shaft. It is sold by all of the industrial supply houses and it is an important feature to have, as it will save the maintenance people a lot of time.}

Bearing Pillow Blocks:

- Heavy or moderate shock Brand X series with four bolt mounting (or approved equal) ductile iron housing

- Light or no shock: Brand Y series with tow bolt mounting (or approved equal) ductile iron housing

Steel Roller Chain Sprockets:

- Sprockets shall be fitted to give a shaft protrusion of ¼"
- Shafts *3 ⁷⁄₁₆"* diameter and less, sprocket with *Brand X split taper bushing.*
- Shafts *3 ¹⁵⁄₁₆"* diameter and over sprocket with straight bore and ASA FN1 interference fit on shaft; two set screws one on key, one at 90° to key.

Sheaves: *Brand X with QD bushing*

- Sheaves shall be fitted to give a shaft protrusion of ¼"

V-Belt: *Standard 3V, 5V, or 8V*

Guideline 17

{All belts should be matched sets, i.e., all belts are the same length. This is generally understood but there are times when you will not get a matched set. They are cut at the plant at the same time. If the set is not matched, you will have an unmatched belt not doing any work, so on a four-belt set you could be using only three belts.}

Shafting:
- Fabricated items designed to incorporate standard shaft stock shall use the following diameters in inches. $^{15}/_{16}$", $1\ ^{7}/_{16}$", $1\ ^{15}/_{16}$", $2\ ^{7}/_{16}$", $2\ ^{15}/_{16}$", $3\ ^{7}/_{16}$", $3\ ^{15}/_{16}$", $4\ ^{7}/_{16}$", $4\ ^{15}/_{16}$", $5\ ^{7}/_{16}$".
- Keyways shall be done to ANSI Standard, square, parallel key, round ends, unless specified otherwise.

Guideline 18

{Shafting used in the plant should be kept standard as most plants keep spare shafting material in stores. Odd size shafting only adds to the plant costs and potential delays.}

Threaded connections shall conform to American standards.

Guideline 19

{You want American standard threaded connections even if the product comes from a metric country. This way the plant does not have to keep a supply of metric fittings in stock. In remote locations you may not be able to obtain metric fittings, so it is best to stay away from them. However, you may not be able to get this requirement through for large, heavy equipment built in Europe.}

Lubrication Grease Fittings: ***Brand Y*** hydraulic fittings (or approved alternative)

Roller Chain: Standard ANSI, steel, riveted construction.

Gear Reducers: Shaft Mounted – ***Brand Y***
                    In Line Units – ***Brand X***
                    Parallel Shaft – ***Brand X***

Bolts and Nuts - ANSI B18.2, Heavy Hexagonal (electro-cadmium plated on structural steel)

Piping

- Standard valves shall have the following connections:
  a) 2" and less, screwed or socket weld.
  b) 2 ½" and up-flanged or butt-welded.
- Hydraulic oil piping, tubing, hoses, and valves shall conform to plant design standard.
- Hydraulic oil and lubrication oil tube fittings, hoses, and valves shall conform to plant design standard.

Guideline 20

{Plant management will have to determine if they want to make the valve connections 2" and less or 2 ½" and less. Check to see what specifications were used for the plant construction.}

Instrumentation

Electronic Instrument Standard Control Signal: 4-20 MA
Pneumatic Instrument Standard Control Signal: ***20-100 kPa (g)***

Plant Standard Instruments and Accessories:

Air Signal Tubing Seamless 316 Stainless Steel.
½" O.D. x 0.035" wall (supply)
½" O.D. x 0.025" wall (supply)

Plant Standard Instruments and Accessories	
Pneumatic Tube Fittings	*Brand X*
Pneumatic Push-Lock Flexible Hose	*Brand Y*
DCS System	*Brand X*
PLC (Programmable Logic Controller)	*Brand Y*
Ball Valves – Throttling	*Brand X*
Butterfly Valves – Throttling	*Brand Y*
Globe Valves – Throttling	*Brand X*
Plug Valves – Throttling	*Brand Y*
Ball Valves - On/Off	*Brand X*
Butterfly Valves – On/Off	*Brand Y*
E - Disc Valve - On/Off	*Brand X*
Globe Valves - On/Off	*Brand Y*
Knife Gate Valves - On/Off	*Brand X*
Plug Valve - On/Off	*Brand Y*
Backpressure Valve	*Brand X*
Pressure Reducing Valve	*Brand Y*
Pressure Sustaining Valve	*Brand X*
Specialty Valve	*Brand Y*
Boiler Fuel Safety Valves	*Brand X*
Damper Drives – Valves Actuator	*Brand Y*
Transmitter Isolation Valves	*Brand X*
Solenoid Valves	*Brand Y*
Transmitting Equalizing Manifold Valve	*Brand X*
Air Regulator Filter (for control valves)	*Brand Y*
Diaphragm Seals	*Brand X*
Analyzer Transmitter (Streaming Current Detector)	*Brand Y*
Consistency Transmitter	*Brand X*
Current Transmitter	*Brand Y*
Gauge Pressure Transmitters	*Brand X*
Flange Mount Transmitter	*Brand Y*
Differential Pressure Transmitter	*Brand X*
Liquid Level Transmitter	*Brand Y*
Magnetic Flow Transmitter	*Brand X*
Position Transmitter	*Brand Y*

Plant Standard Instruments and Accessories (cont.)	
RF Capacitance Type Level/Flow Transmitters, Switch	*Brand X*
Temperature Transmitter	*Brand Y*
Ultrasonic Level Transmitters, Switch	*Brand X*
Vibration Transmitter	*Brand Y*
Alarm Switch	*Brand X*
Flow Switch	*Brand Y*
Flow Switch (RF Capacitance)	*Brand X*
Level Switch	*Brand Y*
Level Switch (Conductance)	*Brand X*
Level Switch (Gamma)	*Brand Y*
Level Switch (for Control Valve)	*Brand X*
Position Switch	*Brand Y*
Pressure Switch	*Brand X*
Proximity Switch	*Brand Y*
Temperature Switch	*Brand X*
Torque Switch	*Brand Y*
Speed Switch	*Brand X*
Mass Flowmeter (Coriolis Type)	*Brand Y*
Rotameters/Purgemeters	*Brand X*
RTD & Thermowell	*Brand Y*
Thermocouple	*Brand X*
Pressure Gauge	*Brand Y*
Temperature Gauge	*Brand X*
Ambient Air & Stack Analyzer	*Brand Y*
Bleach Analyzer	*Brand X*
PH/ORP Analyzer	*Brand Y*
Conductivity Analyzer	*Brand X*
Turbidity Analyzer	*Brand Y*
Nuclear Density Meter/Level Gauge Analyzer	*Brand X*
O_2 Analyzer	*Brand Y*
Closed Circuit TV	*Brand X*
TV Monitor	*Brand Y*
Sequential Video Switcher	*Brand X*
I/P Transducer	*Brand Y*
Linear Position Transducer	*Brand X*

Plant Standard Instruments and Accessories (cont.)	
Speed Transducer	*Brand Y*
Signal Conditioners (I/I, E/I, etc.)	*Brand X*
Refractometer	*Brand Y*
Stack Gas Flow	*Brand X*
Opacity	*Brand Y*
Hazardous Gas Monitor	*Brand X*
Rupture Disc Alarm Monitor	*Brand Y*
Pitot Tubes	*Brand X*
Flow Element - Orifice Plate	*Brand Y*
Metering Pumps	*Brand X*
Ammeter	*Brand Y*
Local Level Indicator	*Brand X*
Grounding Indicator	*Brand Y*

Electrical

Any motors that must be supplied as an integral part of the equipment shall be: three phase, WP1 standard efficiency manufactured by **Brand X** for medium voltage and TEFC, high efficiency, severe duty, manufactured by **Brand Y** for low voltage with Class F insulation, with Class B temperature rise, 1.15 service factor, and suitable for application in the ***pulp and paper*** industry.

Guideline 21

{Change the motor specifications to suit your plant. Check the voltage break for medium and low voltage. This motor specification will give you a very heavy duty motor}.

Operator's control panel devices shall be heavy duty oil-tight type –
Brand X or approved alternative.

- Start stop nameplate: 26 mm x 70 mm (including engraving).
- AC motors electrical load indication, meters with 5-amp full-scale movement and scale calibrated – 0-150% scale.
- Control relays – ***Brand X***
- Terminal blocks – ***Brand Y***

Guideline 22

{The nameplates are called Lamacoids and can be either white engraving on black or black engraving on white. You should specify what your plant is currently using. With meters you have to watch your scales. If the meter has a scale with too wide a range, you will not be able to read the minor movements accurately }.

Any motor starting or controlling equipment, which must be supplied as part of an equipment package, shall be programmable controller compatible.

- Wiring of starter unit shall be to plant standard *0C-11.11*
- Solid state programmable controller: ***Brand X c/w 120 V AC I/O system.***

Guideline 23

{If your plant is using a DCS system check the programmable controller requirement.}

Plant Standard Electrical and Accessories	
Limit Switches	*Brand X*
Pressure Switches	*Brand Y*
Proximity Switches	*Brand X*
Pull Cord Safety Switch	*Brand Y*
Control Transformer	*Brand X*
Current Transformer	*Brand Y*
Potential Transformer	*Brand X*
Power Transformer	*Brand Y*
Motors	*Brand X*
Coupling – Tachometer Generator	*Brand Y*
Programmable Controllers	*Brand X*
Zero Speed Switch	*Brand Y*
Variable Speed Drives	*Brand X*
Motor Protections	*Brand Y*
DC Motors	*Brand X*
Pulp Machine Drives	*Brand Y*
MCC	*Brand X*
Starters, Contactors	*Brand Y*
Medium Voltage (2400 V)	*Brand X*
Timers	*Brand Y*
Photo-cells	*Brand X*
Control Relays	*Brand Y*
Counters	*Brand X*
Relay Sockets	*Brand Y*
Pushbuttons	*Brand X*
Indicating Lights	*Brand Y*
Selector Switches	*Brand X*
Joy Sticks	*Brand Y*
Lighting Panels	*Brand X*
Solenoids	*Brand Y*
Terminal Blocks	*Brand X*
Fire Alarms	*Brand Y*
Cable Trays	*Brand X*
Cable Connectors	*Brand Y*
Electric Heating	*Brand X*

Plant Standard Electrical and Accessories (cont.)	
Welding Outlets	*Brand Y*
Wiring Devices	*Brand X*
Receptacles:-Duplex Grounding Type	*Brand Y*
Welding	*Brand X*
Switches:-Single Pole 15A	*Brand Y*
600 V Switchgear	*Brand X*
13.8 kV Switchgear	*Brand Y*
Control Fuses	*Brand X*
Wire Markers	*Brand Y*
Heat Trace	*Brand X*
Junction Boxes	*Brand Y*
Belt Off Track	*Brand X*
Space Heaters	*Brand Y*
Lighting Fixtures:	
Hi – Bay	*Brand X*
Medium Height	*Brand Y*
Fluorescent Lighting	*Brand X*
Emergency Lighting	*Brand Y*

Painting

For painting specifications, surface preparation, materials, and application, refer to "Plant Mechanical Design Standards."

Plant Mechanical Standards as appropriate.

- Structural Steel & Building.
- For Field Assembled Vessels, Breaching, Ducts & Stacks.
- Painted Equipment.
- For Piping, Electrical & Instrument Supports.

Guideline 24

{Always give a paint specification. Give the contractor or vendor two or three choices of manufacturers, but spell out what type of paint material you want. If the contractor or vendor wants to use his own specification, review the specification and approve or disapprove it. If you do not spell out what you want, you will likely get colored water}.

Vendor Data Requirements

Guideline 25

{The information the vendor supplies to you will be used in the bid evaluation. The information requested has to do with the design. The information is used to see if the equipment is suitable and will it fit in the space assigned. If the equipment does not fit, what additional costs are required to make it fit?}

A. **Information Required with Your Proposal**

  1. **Components**

  In addition to describing fully the proposed equipment, the vendor shall also specify in full detail any of the following components, which may be purchased by him to constitute part of the proposal. These components should conform to the Plant Conditions and Standard Component List *(attached)* and vendor shall advise which components do not conform to the plant standard.

  Pipe & Fittings:

  - Dimensional specification, metallurgy, wall thickness, and method of manufacture.
  - Valves - Manufacturer, model number, type, and metallurgy.

Electrical:

- Motors required for equipment operation (whether included in bid or not), recommended hp, speed, and characteristics.
- Other electrical components: manufacturer, model, type, and rating.
- Gear Reducers - Manufacturer, model number, type, service factor, and A.G.M.A. (American Gear Manufacturers Association) rating.

Guideline 26

{Most plants have purchasing agreements with motor manufacturers and get very good prices from them. The plant will usually purchase the motor separately and, if necessary, have it sent to the vendor for mounting with the equipment.}

2. Proposal Drawings

Drawings should fully illustrate:

- The physical dimensions of equipment specified.
- The necessary clearances around the equipment required for operation and maintenance.
- Special foundation requirements.

Guideline 27

{Requiring dimensions allows you to prepare a layout to see if the equipment will fit. If it won't fit, consider what would be required to make it fit and whether or not this is worth doing. Similarly you want to know about the foundation. A foundation out of the ordinary could result in a higher cost or affect operation of other equipment.}

3. List of Drawings and Other Documents to Be Supplied with an Equipment Order

This list should include the following where applicable:

- General arrangement of all included equipment
- Electrical system schematic
- Instrument system schematic
- Ladder logic drawings for PLCs
- Installation, operating and maintenance manuals
- Complete parts list
- Recommended spare parts list with pricing
- Other pertinent details

Guideline 28

{Make sure you get all drawings from the vendor to enable you to operate and maintain the plant. Before awarding a contract, ask if there any drawings they will not give you. If they will not give you certain drawings for proprietary reasons, consider whether or not you really want to do business with the vendor. If you are missing any information it could potentially cripple the plant.}

4. Access & Field Assembly Data

Details of platforms, stairways, ladders, handrails, and kick plates considered necessary for proper access to the equipment for normal operation and maintenance and a statement as to whether or not they are included in the bid.

Details of the number of components or pieces of equipment that would be delivered to the site, stipulating weight and overall dimensions of each piece, method of assembly (welded, bolted, etc.), and confirmation that the equipment would be shop assembled and match marked before shipping as necessary to ensure a proper fit.

Guideline 29

{The access detail drawings are need for layout purposes to see if the equipment will fit into the given space. Is the access adequate for the plant requirements, especially maintenance? It is important to know the size and weight of components to determine if you can actually handle and get the components into the desired location.}

5. Equipment Weight

Shipping weight, operating new weight, and maximum flooded weight if applicable.

Guideline 30

{You want to know the shipping weights for handling purposes. The flooded weight is the weight of the unit when full of water. This value is used for the structural design.}

6. Transport

The preferred method of transportation. If equipment is too large or heavy for shipment to the plant site by normal non-restricted transportation methods, specify details. Also advise of any special handling, lifting, and storage requirements.

Guideline 31

{If the equipment is large and coming by ocean, the closest port and transportation will have to be looked at. You can also fly equipment to sites with heavy lift aircraft. At different times of the year, especially spring, there are restrictions on highway loading for trucks. If shipping at the wrong time, route and shipping costs can be affected.}

7. Equipment Operational Services

The quantity and rating of all services required to operate the equipment, such as steam, compressed air, oil, gas, water, electricity, etc.

Guideline 32

{This information is required to determine if existing plant services are adequate}.

8. Metallurgy

Full metallurgical detail of the proposed equipment.

9. Erection and Startup Supervision

The daily charge-out rate, including subsistence cost, for qualified field engineering personnel deemed necessary by the manufacturer to supervise erection of their equipment.

The daily charge-out rate for an experienced operating engineer to assist startup and initial operation. State whether or not this service is necessary for guarantee purposes.

B. Performance Required by the Successful Bidder

The successful bidder will be required to perform to the following standards and/or conditions:

1. Codes

Local codes, standards, regulations, and labor agreements having jurisdiction over any part of the work covered in this bid shall be followed.

2. Structural

Fabricated steel components shall have sharp edges rounded and free of burrs, shall be welded in accordance with recognized good practice, and all weld spatter shall be removed.

Structural members composed of back-to-back angles shall not be used in the manufacture of structures or equipment.

Equipment anchor boltholes shall be ¾" minimum anchor bolt diameter. Equipment bolted to frames and structures shall have the bolt heads on top of the connection components so that the bolts will not drop out of the holes whenever a nut works loose.

3. Mechanical

Baseplates shall conform to the following specifications:
- They shall be of rigid construction
- Adequate grout holes shall be provided on the top surface and vent holes in each corner.
- Equipment and drive setting blocks shall be welded to the baseplate and machined as an integral unit to assure the mounting surfaces are level with one another.
- The drive setting surfaces shall be a height that will allow ⅛" shims to be placed under the drive.
- Setting blocks welded to the baseplate shall have continuous seal welds (skip welding shall not be used).
- Baseplates shall be provided to suit the next larger motor frame size. Spacer blocks shall be welded.
- Equipment shall be fastened to the baseplate with cap screws.
- Bolts and nuts shall be American Standard Heavy N.C.2, electro-cadmium plated with hexagonal heads.
- Washers shall be installed under all nuts and under bolt heads that are placed in slotted holes or bear on cast metal or plastic materials.
- Beveled washers shall be used on tapered surfaces, such as structural steel flanges.

Guideline 33

{Your design drawing should specify a surface finish for the baseplate. Do not send out a baseplate drawing without a specified surface finish shown on it. You can purchase test coupons to compare surface finishes to ensure that you get what you asked for.}

The vendor shall supply bearing assemblies, couplings, sheaves, sprockets, V-belts, chains, baseplates, and equipment guards. He shall fit couplings, sheaves, and sprockets to his equipment as well as bore and key seat them to suit the shaft of the drives as required.

Guideline 34

{If you purchase the motor yourself, you may have to ship it to the vendor for fitting of drive components and testing of the equipment.}

4. Piping

- Carbon steel pipe, tube, and fittings for lubrication and hydraulic oil shall be pickled and passivated.
- Butt weld joints on lubrication or hydraulic oil piping shall have the inside of the joints ground to remove any slag.
- Threaded joints on lubrication or hydraulic oil piping shall have all cuttings and metal chips removed prior to assembly.
- Teflon tape shall not be used for threaded joints.
- Flange bolt hole drilling shall conform to ANSI Standard and shall straddle the centerlines.

Guideline 35

{If Teflon tape is not placed on the joint properly, small pieces can break off and plug up fine holes farther down the line.}

5. Instrumentation

- All tubing and piping shall terminate at a manifold.
- All wiring shall terminate at terminal stops.
- Control panel wiring, piping, and tubing shall be factory checked and guaranteed operational.
- Only instrumentation that has been specially developed to suit the equipment shall be included.

6. Electrical

- Devices shall be by recognized acceptable manufacturers.
- Only special motors that form an integral part of the manufacturers' equipment shall be included.

7. Drawing Requirements

Guideline 36

{In pre-computer days, drawings were done on mylar or linen, and sepia copies were made and issued. With CADD, electronic files, and e-mail today it is easier to send an electronic copy. If you use electronic copies, make sure that you get a hard copy as well. Sometimes, the electronic drawings will get scrambled and you will not get a complete drawing. When using electronic drawings you need a color printer or plotter, otherwise the drawings will be in color on the screen and black and white on the plot. It is difficult to follow flowsheets prepared in color that plot as black and white. Make sure you have engineered stamped drawing requirements met by the vendor.}

The following drawing copies are required:

Mechanical Equipment	
Preliminary issue	*2 prints, 1 reproducible*
Certified loading diagrams and foundation bolt hole location plans	*6 prints, 1 reproducible*
All other certified issues	*6 prints, 1 reproducible*
Electrical Apparatus	
Preliminary issues	*2 prints, 1 reproducible*
Certified issues	*6 prints, 1 reproducible*
Structural Assemblies	
Preliminary issues	*2 prints, 1 sepia*
Certified issues	*6 prints, 2 sepias*
Design calculations	*2 Sets*

8. Manuals, Parts Lists, and Spare Parts Quotations

Where the type of equipment necessitates, *six* copies of installation, operation, and maintenance manuals; parts lists; and spare parts recommendations and quotations are to be furnished at least 60 days before shipment of equipment. Maintenance manuals must contain full lubrication instructions, and the parts list should stipulate the manufacture's name and model number of all purchased components. If equipment is delivered to the site in component pieces, the installation instructions are to contain drawings detailing erection marking, etc.

Guideline 37

{If you require more than six copies, ask for them now, as they will cost you money later. With spare parts, the vendors will usually give you their part number and not the manufacturers', as they want you to purchase spare parts from them. At the first opportunity, find out who the manufacturers are or find an industrial supply house that can provide what you require.}

9. Craftsmanship

The workmanship and material used in the equipment offered must be new and of the highest quality in every respect, and proper regard must be paid to duplication and interchangeably of parts.

10. Shop Testing, Inspection, and Progress Reporting

Where the type of equipment necessitates, the goods furnished shall be shop tested and the purchaser's inspector shall, at purchaser's option, have the right to witness testing. A certificate of the shop test shall be furnished to the purchaser in any event.

Authorized representatives of the purchaser shall be allowed access to the vendor's shops during manufacture to inspect the testing of the goods or parts and to obtain information as to the progress of the work.

Major Equipment

If requested by the purchaser, the vendor shall furnish within six weeks from order date a planned program detailing when the manufacturing functions, including engineering, procurement, etc., are expected to start and end for the various equipment components. Thereafter, monthly reports showing the actual progress are to be furnished until the equipment is shipped complete.

Project Committed Equipment

On multiple items that are generally committed for the project and thereafter ordered individually as the exact needs become known, the vendor shall furnish within six weeks of the first item ordered a report organized by PO number and listing the purchaser's identification number, date received, and expected shipping schedule. The report is to be updated monthly adding new ordered items and indicating shipping dates or revised schedules as applicable. Before first issuance the format shall be reviewed with the purchaser.

11. Tagging of Equipment

Each item shall be tagged with the equipment number and other information, which will be shown in the purchase order. Tagging materials shall be weatherproof and securely attached to each item.

Shipping tag shall consist of:

TAG: (Generic name of item) _____

Equipment No.   _____

PO Item No.   _____

PO No.   _____

In addition, a stainless steel nameplate stamped or engraved with the equipment number shall be attached with rivets or screws directly to the piece of equipment.

Electric motors will have the plant maintenance number on a separate stainless steel plate.

Instrumentation items, small valves, etc. will have the stainless steel plate securely wired to the item.

Vendor Information Requirements

Guideline 38

{The Vendor Information Requirements form should be sent out with RFQs for equipment supply. The vendor should fill it out and return it with the bid. This form requests information of interest to you. The questions are important as the wrong answer could disqualify the bidder or force the plant to work with a supplier when it is not to their advantage to do so.}

Vendor Information Requirements Form

The Vendor must complete this form and attach it to his proposal, supplementing information shown under the "Vendor Furnished Data" heading in his Proposal where applicable. Failure to comply with this requirement may, at the option of the purchaser, disqualify tender.

Description	**Vendor Furnished Data**

I   Drawing and Manuals

a)   Location of Engineering & Design Office

_____

_____

_____

_____

_____

_____

_____

_____

_____

b)   Number of days required after receiving Purchase Order
     Number for submission of:

i	Certified loading diagrams	_____ days
ii	Foundation bolt hole location plans	_____ days
iii	Preliminary mechanical layouts.	_____ days
iv	Certified Drawings (allow two weeks for our approval of above)	_____ days
v	Listing of major spares.	_____ days
vi	Operation & Maintenance Manuals c/w priced operational recommended spare parts list.	_____ days

II Production & Delivery

a) Name of company submitting bid:

_____

b) Equipment manufacturer:

_____

_____

_____

c) Location of manufacturing plant:

_____

_____

_____

d) Name and address of office/distributor closest to Plant:

_____

_____

_____

_____

e) Number of days after receiving Purchase Order
Number for submission of production schedule. _____ days

f) Shipment from plant _____ days

g) Method of shipping _____

h) Maximum lift required _____

i) Maximum unit dimensions   - length _____
                                 - width _____
                                 - height _____

j) Delivery to millsite _____ days

III Environmental Data

a) State the sound level which proposed equipment
will produce if in excess of 85 dB _____ dB

b) Indicate proposal page number where it is stated
that goods comply with: (1 or 2, whichever is
applicable)

   1) The Occupational Safety & Health Act Standard
      (OHSA) _____ page

   2) *Provincial* Department of
      Labor Health & Safety Regulations. _____ page

Plant Project Engineering Guidebook

Contracts and Tender Documents

Normally, a plant will buy its own equipment and then hire a contractor to install it. When using this type of contractor service, a contract of some form should be entered into. A construction contract may take the following forms:

- A purchase order with the General Conditions For Site Contractors attached;
- A construction contract with a detailed tender document containing a Construction Agreement, General Conditions, and Special Conditions.

Your plant should have a standard tender document—ideally one where you can fill in the blanks with the required information. The document is prepared as part of a Request for Quotation and forms the backup to the construction contract upon the award of the work to the successful bidder. The next chapter will take you through a complete tender document.

When you use a construction contract with a tender document versus a purchase order with General Conditions is not a clear cut decision. Your decision should be based on your perception of how complex the work is, the level of risk to the plant of something going wrong, and the dollar amount of the work involved. As these factors increase, it is wise to consider the construction contract with a tender document as it provides more protection. As a guideline, the construction contract with a tender document should be considered for contracts over $100,000 in value; however, using a purchase order with general conditions for contracts greater than $100,000 but involving little complexity and low risk would be acceptable. On the other hand, for contracts less than $100,000 but with increased risk, the construction contract with a tender document may be the better choice. In any situation involving on site labor, either one of these documents should be used.

One other way of getting around this issue is to add your new contract to an existing contract. If you have a contractor working on your site who is working under a construction contract with a tender document, you can issue a purchase order with general conditions stating that "all the conditions of Contract XYZ apply to this purchase order."

The use of the purchase order with general conditions for small to medium sized design/build (EPC) and supply/erect contracts is suitable, as the tender document does not fit these applications very well. Major design/build and supply/erect contracts should be written specifically for the work involved.

Before preparing and sending the tender document, determine what type of contract you will be using. Examples of current types of construction contracts are:

1. **Lump Sum Contract or a Stipulated Price Contract.** In this type of contract the contractor will give you one price for all of the work. The contractor will have included money for his perceived risk. For you to get the best price, your scope has to be very well defined. If not, you will run into the problem of extras. In these types of contracts you do not have much say in how the work is to be carried out. If there are special construction requirements, they should be spelled out in the Scope of Work. If you try to change the requirements after contract award, it could result in an extra charge. Resist all attempts to make changes. Based on your scope, the contractor will determine how long he will be on site and include this in his costs accordingly. Any delays caused by the plant could result in an extra charge to the contract. Get unit prices to cover potential extras to the contract and you should get time and material rates for stoppages due to plant problems. Use this type of contract only when your scope is well defined. There is the potential for lawsuits with this type of contract. The Lump Sum Turnkey or Engineer, Procure, and Construct (EPC) contracts are of this form (see item 7).

2. **Unit Price Contract.** This type of contract requires the contractor to quote prices for individual pieces of work, e.g., price per foot for 2" pipe or price per yard of excavation, etc. This requires determining (or guessing well) what quantities will be involved in the work. For excavation there will be different costs for different types of material excavated, depth of excavation (e.g., one price for 1' to 10' and another price for deeper than 10'), etc. For concrete work there will be different prices for the different types of structures. Talk to the contractors about quantity pricing before preparing your tender package. These contracts can be difficult to manage in plants since the operations usually take precedence over construction, and it is not unusual to have to kick the contractor out of an area negating his unit price quote. To be prepared get pricing for time and material work. With a Unit Price Contract there is an incentive for the contractor to get the work completed as soon as possible.

3. **Cost Plus Percentage (or Time and Material) Contract.** This type of contract requires the contractor to submit labor rates for the various trades as well as his percentage add-on for overhead and profit. The labor rates and benefits are usually fixed by labor agreements; however, the percentages for overhead and profit are negotiable. You will have to keep track of hours spent on daily basis by signing time sheets. It is difficult to estimate the costs of this type of contract. You may be able to make a good guess at the labor required, but you will probably miss on materials. You will authorize obscure items that you would never have thought of as being required. With this contract the contractor makes his money by keeping the costs up. There is no incentive to reduce the costs or complete the work sooner. Keep detailed accounts of everything the contractor submits for payment. This type of contract can be used on large jobs, depending on your company philosophy. Some people feel that by the time you add the costs for the delays, lawsuits, risk, etc., this type of contract is less expensive than a lump sum contract.

4. **Cost Plus Lump Sum Fee Contract.** This contract is similar to the cost plus percentage, but the contractor receives a fixed fee based on an agreed amount rather than on a percentage for overhead and profit. For this you will have to agree on a scope of work, and any additional work will require a corresponding increase in the fee. You can also have a clause that the scope has to increase by more than a certain percentage before there is an increase in the fee.

5. **Cost Plus Lump Sum Fee Plus Bonus Contract.** With this contract the contractor receives a fixed fee based on a defined scope but will receive an additional bonus based on agreed upon criteria. The bonus could be a sharing of any construction savings or a sharing of increased revenue from the increased production. The savings should come from his management of the work, not from supplying substandard equipment or parts. Along with the bonus there could be a penalty clause. Here the contractor would pay a penalty if he did not meet schedule criteria spelled out at the beginning. You cannot have a contract with only a penalty clause; the contract has to have a bonus and a penalty clause.

6. **Guaranteed Maximum Price Plus Bonus Contract.** This type of contract gives the contractor an incentive to save money. The guaranteed maximum price should be based on detailed plans and specifications and not on a cost estimate. The owner and contractor can split any savings and the contractor assumes any costs over the guaranteed maximum price.

7. **Engineer, Procure, Construct (EPC) or Lump Sum Turnkey Contract.** In today's (2000) atmosphere, the EPC contract is the contract of choice for large projects. With this contract the contractor does all of the engineering design, procurement, and construction for one agreed upon price. The contractor takes all the risk, but by the same token, his price will include a fee to cover this risk. You will only get one total price from the contractor. Do not try to figure out what his area costs are from the progress payments as they will not be a true reflection of his costs. To get what equipment you want you should spell out design parameters

and the desired equipment manufacturers during the bidding phase. Once the contract is awarded, you will get whatever the contractor used for pricing and will have no opportunity to change it without a cost penalty to you. This type of contract is also prone to lawsuits. There can be great risk to the plant if the contractor is wrong and goes bankrupt halfway through the project. These contracts can also contain liquidated damages. These are penalties the contractor must pay if he does not meet certain criteria such as schedule dates and process guarantees. This type of contract has been the ruin of several major construction companies within the past few years.

You can purchase, from organizations such as the Canadian Construction Document Committee (CCDC), fill-in-the-blank construction contracts. These contracts are used in Canada by owners, engineers, contractors, and architects. CCDC2, which is referred to in the tender document in the next chapter, is the standard Lump Sum Contract. When your company policy differs from the CCDC standard contract, you will have to modify the CCDC contract through the Special Conditions section in your tender document. The sections that may differ relate to insurance and bonding. The tender document to be reviewed next contains these types of changes. As with all legal documents, you should get legal advice before using any of these standard documents.

CCDC publishes a variety of useful materials that relate to contracts. Contact them for a list of their publications at:

> Canadian Construction Documents Committee
> 85 Albert St.
> Ottawa, Ontario, Canada, K1P 6A4
> http://www.csc-dcc.ca/publicat/index.html

Now that you have determined which type of contract you are going to use, you can now start work on assembling your tender document for issue as part of a RFQ.

TENDER FORM AND INSTRUCTIONS TO BIDDERS

The tender document discussed here is for a Lump Sum contract and was developed for use in British Columbia, Canada. Treat it as a guide only and modify sections to suit your situation. As with all legal documents, you should get legal advice before using any of the wordings included here. With modifications, this document can also be used for all other types of contracts. You will have to review and change what does not apply to your particular contract. Your tender document should contain all the forms you want the contractor to submit and all the information you want him to track.

With word processing, it is simple to store this tender document once you have modified it for your plant. When you need to make up an RFQ, it is a matter of filling in the blanks. The footer should have the RFQ number in the bottom right hand corner to identify the documents. Make sure you change this number for each RFQ. Do not issue an RFQ with different numbers on the pages.

If you require more than one copy of the bid returned from the vendor/contractor, the tender document or RFQ should state how many copies are required to save you making copies later. It is easier to get the vendor to send the required number of copies with his bid. All bids should be sent to the Purchasing Department where the time of arrival and the bidder will be recorded. They will also follow up on bids not received by the specified deadline. Purchasing should open all quotes with or without your presence. Do not open bids yourself or before the appointed time given in the tender document as this can lead to ethical questions.

All dealings with contractors should be above board and professional in every way. All correspondence during the bid period should be copied to the Purchasing Department. It is important that the Purchasing Department have a complete file of all correspondence as it all forms part of the contract or purchase order that will be written. Purchasing should be the one source of all information related to the bidding process, so keep them apprised.

With modern communications it is now normal for contractors/vendors to fax their bids by the closing date and send the original document by courier. Unless you have a secure fax machine, i.e., one that is not accessible to the general plant population, the faxes should be sent to the Purchasing Department. Normally faxes sent to engineering departments are left lying around the fax machine for everybody to read. This is not desirable as the bids are confidential information and should be treated as such. For the information you require, get a copy from the Purchasing Department leaving them the original.

The tender document is broken down into six sections. Section I is the Tender form, which the contractor is to complete and return to you. This is the document with his dollar value on it. The other five sections of the tender document are the backup material for the bid. The contractor keeps these sections as reference material.

Normally all technical questions are answered by you and questions regarding terms and conditions are answered by your Purchasing Department. All RFQs should state this, and it is important to include names, phone numbers, and fax numbers of the relevant contacts.

If, during the bidding process, a contractor has questions about the scope and how to do the work, you will have to decide if the other bidders should know the answer. If a contractor asks a question that will give him a competitive advantage, the answer should not go to all contractors. This situation usually comes about from the contractor coming up with a less expensive way of installation and the other contractors should not benefit from his knowledge. Generally answers to questions regarding technical clarifications should be distributed to all bidders.

The progress payment forms are a standard format used throughout the construction industry. They were developed for use in Canada, so provincial sales tax (PST) and the goods & services tax (GST) is added on at the bottom of the form. You can use these numbers, change them, or delete them, depending on what the tax regulations are in your jurisdiction. The same applies to the holdback. You should prepare these documents in a spreadsheet format for automatic calculation.

The following numbered items correspond to the blanks in the tender document that have to be filled in.

1. Title of the project.
2. Project authorization number.
3. Request for Quotation number issued by purchasing. This number should also be put in the footer so it appears in the bottom right hand corner of every page.
4. Date bids are due
5. Name of the contractor to whom you are sending the Request for Quotation.
6. Contractor's full mailing address. This address may not be the mailing address later if he is the successful bidder, as he will set up on site. For legal issues, this should be the address where documents are to be sent.
7. The contractor enters his total value here. This value should be written out in words and a number value entered. Check to make sure they match. If they don't match, get the contractor to change it and resubmit the front sheet. Do not destroy the sheet that is incorrect, but file it with the RFQ.
8. The contractor fills this in. In Canada there is a goods & services tax. It may not apply in your jurisdiction. You can leave it out or put in the name of the tax that applies to you.

9.   The contractor enters the total cost for his labor. If you subtract 8 and 9 from 7 you should get a rough idea of what the contractor's material and equipment costs are.

Item nos. 10 to 16 refer to the Statement Regarding Outstanding Claims. This form was developed to get around the problem of a contractor submitting a claim months after the work is finished and to allow you to keep track of extra charges. The contractor should be told before contract award that this form has to be submitted with his progress payments, and he can not submit a claim unless it has been identified on this form. This form is left blank in the tender document for the successful contractor's use after contract award. The contractor should fill in all the information:

10. Contract number
11. Contractor's company name
12. Contract title
13. Date contract was issued
14. Invoice date
15. Contractor's signature
16. Company position or title of the person signing the document
17. Witness to the signature
18. Area for the Additions (ADDS) and Deletions (DELETES) you want to apply to the contract. There are times when you will want to add extra work to your scope and other times when you will want to delete items from your scope. To keep control of your costs, you ask the contractor, at the bid stage, for unit prices to cover the items you may want to add (additions) or delete (deletions) at a later date. Examples would be earth works, concrete in foundations or walls, steel, valves, piping, pipe fittings, cable, terminations, etc. If you do not ask for these unit prices before contract award and have them written into the contract, you are at the mercy of the contractor to provide reasonable unit prices later. List all items for which you require a unit price. These are items you feel are not well defined in the scope. You should also make a list of Deletes. You will receive credit for deducting these

items from the scope. The Delete value will always be less than the Additions. The list can be as long as needed. If you are doing concrete work, make sure you clearly define whether or not the price includes rebar and formwork. Concrete prices will be different for walls than for slabs on grade, so you will have to look at all the types of concrete work you have. Excavation prices will be different based on depths of excavation, material excavated, location of excavation, etc., so look at what excavations are required. Steel is divided into miscellaneous, small structural, and large structural steel. The miscellaneous steel has to be well defined as it can become confusing otherwise. When putting down an item for Additions, take a close look at how the contractor prices the item and make sure you have it covered completely. Contractors will tell you what the breakdown on sizes should be. You can discuss these breakdowns with the contractors at bid meetings.

19. In case you have missed something, you want to know what the Cost Plus Percentage will be. This is used when you have a job to do that is not covered by the contract or the unit prices. The labor is covered somewhere else so this is for material and equipment. The contractor fills in this percentage, which is usually negotiable.

20. The percentage for equipment may not be the same as the materials percentage. The contractor fills in this percentage and the percentage is usually negotiable.

21. This is the cost plus for your contractor's subcontractors' material. The contractor fills in this percentage which may be negotiable.

22. This is the cost plus for your contractor's subcontractors' equipment. The contractor fills in this percentage which may be negotiable.

23. For this section make sure that the small tool value is correct. It may have to be higher—your contractors can advise you on the value to use. The rates given should cover all the people that will be on site and should include all costs associated with each trade. You want to know what the rate is for preparing estimates and calculating the total rate. It also

helps you confirm charges when checking invoices. For overtime, the rate should NOT be double the straight time rate. If it is, the contractor is cheating you because he is doubling his overhead and profit as well. The contractor's base rate includes his overhead and profit added to the labor rate including burdens. For overtime, the labor rate changes by an agreed amount, for example 1-1/2 times. To this overtime rate is added overhead and profit. This is a smaller number than 1-1/2 times the straight time rate.

Example:		
Labor		$25.00 / hr
Burdens		$15.00 / hr
Total		$40.00 / hr
OH & Profit @ 15%		$6.00
Total Straight Time Rate		$46.00
Overtime @ 1 ½ (1 ½ x 25)		$37.50 / hr
Burdens		$15.00 / hr
OH & Profit @ 15%		$6.00
Total Overtime Rate		$58.50
$46.00 x 1.5 = $69.00		

In the above example $58.50 - $46.00 = $12.50. The $12.50 is the premium portion of the overtime rate. So when the form is filled in by the contractor, you should see a basic rate/hour of $46.00 and a premium portion of overtime rates of $12.50. Add the two to get an overtime rate of $58.50.

Contractors, upon request, will give you their rate breakdowns showing what the base rate is and what makes up the burdens. They will not show you what their overhead and profit is. The contractor should not add overhead and profit to travel and living out allowances.

24. The first addendum issued - #1.
25. The last addendum issued. These two will give a range for the addendum.
26. The period the contractor has to submit an acceptable schedule to you—usually two weeks, unless you want it sooner. Depending on the complexity of the project, it may take longer than two weeks to generate.
27. Milestone dates are the dates you want the contractor to meet, for example:

Foundations Complete – 31 August 2001

The contractor will use all your milestone dates to develop a complete schedule.

28. The contractor should indicate what equipment he intends to bring to the site. You should review the list, as there may be a piece of equipment that cannot be used on site. An example would be a crane that is too high to get under overhead cables. Make sure the rates are complete including operator, maintenance, and insurance.
29. The contractor should inform you who he will be using as subcontractors. These subcontractors are subject to your approval.
30. An official of the contractor who can confirm that someone in the contractor's employ visited the site in relation to this bid package. If no one visited the site, that should be stated.
31. The name of the person(s) who visited the site in relation to this bid package.
32. The date the person(s) visited the site—usually the date of the site bid meeting.
33. The name of the company submitting the bid.
34. The signature of the company official authorized to sign the bid.
35. The signatory's official title.
36. The date the bid was signed by the above person.
37. The number of sets of drawings issued to the contractor–usually one set.

38. The number of copies of the specifications issued to the contractor–usually one set.

39. The time of day the bids should be submitted to you by. Change the time zone to suit your location.

40. The date you want the bids submitted.

41. The Request for Quotation (RFQ) number. When the bids come into the office they are not opened right away, but are held until the closing time has passed. You need the RFQ number on the bid package so the purchasing group can collect them without opening to see what it is.

42. The name of the person in your plant who should receive the bids—usually the plant purchasing agent.

43. The day of the week the site meeting for contractors will be held.

44. The date the contractor site meeting will be held.

45. The month the site meeting will be held.

46. The time of day when the site meeting will be held.

47. The name of the person who can answer design and technical information. This should be you, as you want to know what is said. A record of the discussion should be made and form part of the contract if necessary.

48. The name of the contact person in purchasing who can discuss terms and conditions. A record should be made of any discussions.

49. Indicate if the plant is providing an office and/or lunchroom for the contractor. If not, enter "NONE PROVIDED." The use of your company's cafeteria should be mentioned here. Make sure this section agrees with the Scope of Work Special Conditions. Generally, contractors are expected to supply their own office trailers and lunchroom trailers. You should also advise contractors that their trailers should be heated with propane and not electricity. Power would be for lighting only. The trailers will be located in a designated area that you will determine, keeping in mind you will have to get power to the trailer.

50. Indicate if the plant is providing washroom facilities for the contractor. If so, describe what and where. If not, enter "NONE PROVIDED." Make sure this section agrees with the

Scope of Work Special Conditions. If the contractor is providing the washcar trailers, they should be propane heated and have either a septic field or holding tanks. They should not be connected to the plant sewer system. You can run a potable water line to the location of the wash cars.

51. Indicate what and where the parking facilities are for the contractor. Indicate if this is winterized parking or not. State whether he can park in the plant employees' parking lot or if is there a separate location for the contractors. Does your plant have a separate entrance for contractors? Make sure this section agrees with the Scope of Work Special Conditions. Winterized parking will require plug-ins for the car block heaters. Also state if the plant will plow snow from the lots and construction areas. It is better if you leave snow removal of these areas to the contractor.

52. Indicate what type of power is provided for the contractor's use. State if the contractor is to provide his own power panels, which is the preferred method. Advise the contractor of the type of power connection available for him. (e.g., Can he use a welding outlet? Can he tie into an MCC?) Advise who ties the contractor's power cable into the plant system. The plant electricians usually do this tie-in. For large projects, a separate transformer is usually installed by the plant to provide construction power. From this transformer power cables are run to panels located at various areas in the plant for the contractor's use. The contractor has his own power panel and hooks up to the connection. Keep in mind that you may have three or four contractors working in one area at the same time. Do NOT give the responsibility of providing construction power to a contractor. If you want to bill contractors for power, you should look at prorating it based on the number of people on site. If no power is to be provided, enter "NONE PROVIDED." Make sure this section agrees with the Scope of Work Special Conditions.

53. Indicate location and connection size for domestic and potable water. Advise if the contractor can use fire hydrants and hoses. Normally, you do not want them using plant equipment, and the use of a fire hydrant may set off an alarm.

If nothing is available, enter "NONE PROVIDED." Make sure this section agrees with the Scope of Work Special Conditions.

54. Indicate what temporary lighting is available from the plant. This usually means extension cords with pigtails and lightbulbs. Don't get involved with supplying temporary lighting—let the contractor provide it. If none is to be provided, enter "NONE PROVIDED. ONLY LIGHTING AVAILABLE IS GENERAL PLANT LIGHTING." Make sure this section agrees with the Scope of Work Special Conditions.

MHS Engineering Services Inc.
P.O. Box 600
Anywhere, B.C.
L5R 3T9

Lump Sum Tender

List of Tender Documents

Section	Title
I	Tender Form
II	Instructions to Bidders
III	Scope of Work
IV	Drawing List
V	Specifications and Standards
VI	Agreement & General Conditions of Contract

Section I

Lump Sum Price
Tender Form

MHS Engineering Services Inc.
P.O. Box 600
Anywhere, B.C.
L5R 3T9

Project Title: _____ 1 _____

P.A. No: _____ 2 _____

Inquiry No: _____ 3 _____   **Closing Date:** ____ 4 ____

Name of Contractor: _____ 5 _____

Address of Contractor: _____ 6 _____
_____
_____

1.0 Having carefully examined the tender documents and drawings listed herein for the above titled project, we propose to furnish all materials, supervision, labor, plant, equipment, and tools required by and in accordance with the said documents for the entire work including all duties, municipal, and federal taxes, for the lump sum of:

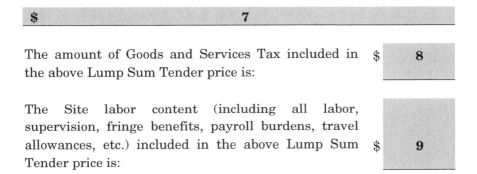

$ 7

The amount of Goods and Services Tax included in the above Lump Sum Tender price is: $ 8

The Site labor content (including all labor, supervision, fringe benefits, payroll burdens, travel allowances, etc.) included in the above Lump Sum Tender price is: $ 9

1.1 Cost and Billing

1.1.1 Billing Procedure and Cost Breakdown

It is understood no advance payments will be made. Progress payments will be made as the work progresses on the basis of Monthly Progress Invoices submitted (see attached form) to the Project Engineer. The applicable holdback as required by law will be released upon acceptance of the work by the Owner.

The Lump Sum price quoted in Section 1.0 is broken down as specified on the attached Progress Invoice form. The breakdown of the total price is required to assist in comparison of bids and for accounting purposes only.

Contract Change Notifications and approved Field Work Orders will be submitted on separate forms provided by the Owner when required.

Owner: _____

Contractor: _____

Project: _____

Contract No.: _____

Contract Name: _____

Progress Invoice No.: _____

Period Ending: _____

Code	Description	Contract Amount	Percent Complete	Value Completed	Previously Claimed	This Claim	Balance
	Sub-Totals						
	Approved Change Notification						
	Approved Field Work Orders						
	Totals						

Less 15% Holdback

Payment Requested This Claim

PST 8%

GST 7%

Total

Approvals For % Complete

MHS Engineering Services Inc.

Contractor

Owner: _MHS Engineering Services Inc._
P.O. Box 8000
Anywhere, B.C.

Contractor: _A.B.C. Name LTD._

Contract No.: _MHS PO Number_

Progress Invoice No.:: _1_

Period Ending: _30 March 2001_

Project: _Building Steel_

Contract Name: _Platforms_

Code	Description	Contract Amount	Percent Complete	Value Completed	Previously Claimed	This Claim	Balance
According to Contract Breakdown e.g. 170	Platforms	200,000.00	10%	20,000.00		20,000.00	180,000.00
Etc.							
List Full Contract Code Breakdown	List Full Contract Description Breakdown						
Sub-Totals		200,000.00		20,000.00	0.0	20,000.00	
Approved Change Notification		30,000.00		15,000.00	10,000.00	5,000.00	**15,000.00**
Approved Field Work Orders		1,000.00		800.00	400.00	400.00	**200.00**
Totals		231,000.00		35,800.00	10,400.00	25,400.00	**15,200.00**
	Less 15% Holdback					3810.00	
	Payment Requested This Claim					21,590.00	
			PST 8%			1727.20	
			GST 7%			1,511.30	
			Total			24828.50	

Approvals For % Complete

MHS Engineering Services Inc.

Contractor

Plant Project Engineering Guidebook

Attachment to Progress Invoice No.:

Contractor:

Period Ending:

Contract No.:

--- List of Approved Field Work Orders ---

Code	Field Work Order No.	Approved Amount	Percent Complete	Value Completed	Previously Claimed	This Claim	Balance
Sub-Totals							

MHS Engineering Services Inc.

Attachment to Progress Invoice No.: 1   **Period Ending:** _30 March 2001_

Contractor: _A.B.C. NAME LTD_   **Contract No.::** _MHS PO Number_

--- **List of Approved Field Work Orders** ---

Code	Field Work Order No.	Approved Amount	Percent Complete	Value Completed	Previously Claimed	This Claim	Balance
170	002	1000	80	800	400	400	200
Sub-Totals		1000		800	400	400	200

MHS Engineering Services Inc.

Attachment to Progress Invoice No.: _____

Period Ending: _____

Contractor: _____

Contract No.:: _____

--- List of Approved Contract Change Notifications ---

Code	Change No.	Approved Amount	Percent Complete	Value Completed	Previously Claimed	This Claim	Balance

MHS Engineering Services Inc.

Attachment to Progress Invoice No.:   **Period Ending:**   *30 March 2001*

Contractor: *A.B.C. NAME LTD*   1   **Contract No.::**   *MHS PO Number*

--- List of Approved Contract Change Notifications ---

Code	Change No.	Approved Amount	Percent Complete	Value Completed	Previously Claimed	This Claim	Balance
As on Change Order							
170	001	30000.00	50	15000.00	10000.00	5000.00	15000.00
		30000.00		**15000.00**	**10000.00**	**5000.00**	**15000.00**

MHS Engineering Services Inc.

Guideline 39

{For advance payments you may get asked to pay for site mobilization, which is acceptable. You should develop the progress payment breakdowns (numbers and description) and fill in the Code and Description columns on the Progress Payment form. The breakdown should match your budget numbers. If you prepare these forms in a spreadsheet, you can give the contractor a blank form and it will calculate the numbers automatically. Check with your accounting group, as they may want one invoice for each billing period with the completed progress payment forms as backup.

The other two forms are for field work orders and for change orders. Only approved dollar values should be entered in to these documents and, if used, the dollar amounts carry forward onto the front form.}

1.1.2   Statement Regarding Outstanding Claims

A copy of this form will be completed and submitted with each progress billing. It is understood the progress billing will not be considered completed and payment will not be made without this document (see blank form on next page).

Statement Regarding Outstanding Claims

(This form is to be completed and submitted with each Contract progress billing. Progress billing will not be considered completed and payment will not be made without this document.)

In the matter of Contract No.	**10**
Between MHS ENGINEERING SERVICES INC. and	**11**
For: (Contract Title)	**12**
Dated: (Date of Contract)	**13**
And in the matter of a Request for Contract Payment	
Work performed under this Contract to Date:	**14**

I/We hereby certify that:

☐   There are no known outstanding claims for work performed beyond the Scope of the Contract, which have NOT been communicated to MHS Engineering Services Inc. in writing.

☐   There are outstanding claims, which have NOT been communicated to MHS Engineering Services Inc. in writing, as noted on the reverse side of this Statement.

☐   There are outstanding claims that have been communicated to MHS Engineering Services Inc., but for which Contract Change Orders have NOT yet been received, as noted on the reverse side of this statement.

15

(signature)

17   **16**

(witnessed)   (position)

(To be signed by an officer of the Contractor)
(Attached additional initialed pages if required)

1.2 Changes to the Work

The undersigned agrees that he will perform additional work in accordance with the terms and conditions as detailed in General Conditions of Contract Part 6 of the Canadian Standard Construction Document CCDC No. 2 (not supplied).

Guideline 40

{Reference any other contract you intend to use.}

Unit Prices

The following unit prices shall apply to changes to the Work whether additions or deletions, and in all cases shall include the supply and installation of materials, supervision, labor, tools, equipment, consumable, expendables, overhead, and profit.

18

1.3 Cost Plus Percentage

a) The cost plus percentages for changes to the Work or for extra work by the Contractor and not covered by unit prices:

(1) For Materials:

19

(2) For Equipment (operated):

20

b) cost plus percentage for changes to the Work or for extra work by the subcontractor and not covered by unit prices:

(1) For Materials:

21

(2) For Equipment (operated)

22

Note: Neither these nor any other percentages shall be applied to the unit prices listed in 1.2.

1.4   Labor Charge-Out Rates

The following labor charge-out rates shall apply when work is being performed on a cost plus basis and shall be all inclusive of payroll burdens, overhead, profit and supervision, small tools, expendables, and consumables.   Small tools are tools having an individual purchase price of $500.00 or less.

Class	Basic Rate/Hour	Premium Portion of Overtime Rates

Guideline 41

{Check with contractors in your area to see what value they would consider for small tools.}

23

Note: The rates shall include for travel and living allowance. These will be in accordance with local labor agreements if applicable. No mark-up will be allowed on these.

1.5   Addenda

Addenda No. _____24_____ to _____25_____ inclusive has/have been received, and the Tender price includes a sum for all the Work as called for and/or implied by the said addendum/addenda.

1.6   Schedule

1.6.1   Time is to be considered the essence of this Contract; completion dates given below are tied to Plant operations, and overrunning these dates will result in loss of revenue by the Owner. The Owner may seek reimbursement of such losses should the Contractor be responsible, in whole or in part, for not achieving the contract schedule.

1.6.2   The Contractor agrees to submit a bar chart schedule acceptable to the Owner within _____26_____ days of contract award. Such bar chart shall contain sufficient detail to clearly itemize the major items of work to be done under this Contract. Milestone dates have been included below and overall dates shown thereon are to be adhered to by this Contractor.

1.6.3   The Contractor will attend weekly Site meetings to coordinate and review the work progress. For this purpose two- week schedules of proposed work will be submitted for review at these meetings.

1.6.4   Contractor shall complete a Manpower Report and submit it to the Project Engineer daily (see attached sheet).

<u>**Milestone Dates**</u>

27

Contractors Daily Manpower Report

The information required on this form should be completed daily and returned no later than 9:00 a.m. the next ordinary working day to the receptionist in the main office. You need this form or some version of it submitted every day as outlined.

Contractor (including subcontractors):_____		Contract:_____ Date:_____				
Union Local (If Applic.)	Trade	Foreman on Payroll	Staff on Payroll	Total No. on Payroll	No. Worked Today	Hours Worked Today
	1. Boilermakers					
	2. Boilermaker welders					
	3. Bricklayers					
	4. Culinary workers					
	5. Carpenters					
	6. Cement masons					
	7. Electricians					
	8. Glaziers					
	9. Insulators					
	10. Ironworkers – Reinforcing					
	11. Ironworkers – Structural					
	12. Laborers					
	13. Millwrights					
	14. Operating engineers					
	15. Painters					
	16. Pipefitters					
	17. Pipefitter welders					
	18. Sheet metal workers					
	19. Teamsters					
	20. Other union					
	Sub Total					
	21. Non-trade (manage. super.)					
	22. Administrative & clerical					
	Total					
	Additional Information					
	How many above are:					
	Local Hires - Union Members					
	Local Hires - Permit					
	In-Province Hires					
Subcontractors for this contract:						

_____
Signature

1.7   Equipment

The equipment, which the Contractor shall bring to the Site for the performance of the Work, will be as follows:

Description of Equipment   **Number of Pieces**

28

The Contractor is to supply a schedule of equipment rental rates that will apply to this Contract. This should include daily, weekly, and monthly rates for non-operated equipment and hourly, daily, weekly, and monthly rates for operated equipment. This schedule is to include all applicable costs including operator, maintenance, and insurance.

1.8   List of Proposed Subcontractors

The following are the names and business address of all subcontractors who may perform work or render services to the Bidder in or about the work, together with a statement of the portion of the work. It being understood that no subcontractor is entitled to subcontract or assign, in any way, any portion of his work without the prior approval of the Owner and/or Owner's Representative.

Name/Address	Portion of Work	Union Affiliation	Union Expiration Date
		29	

1.9   Site Supervision

Tenderers shall attach to their Tenders an organization chart indicating the names and positions held by the following:

1. Officers of the Company.

2. Person at Head Office who will be responsible for this Contract.

3. Superintendent to be in charge of the Work at the Site.

4. Other supervisory personnel proposed at the Site.

The Contractor will herewith confirm that he has visited the Site and is fully conversant with all Site conditions.

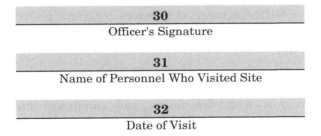

30
Officer's Signature

31
Name of Personnel Who Visited Site

32
Date of Visit

1.10   Tender Information Required

Each Tender shall comprise all of the following:

*   Completed Tender Form, Section 1 items 1.0 through to 1.11.

*   An outline of the manner and sequence in which it is proposed to perform the work, the types and number of various equipment to be used, the length of the work week and any overtime or shift work planned to accomplish the Work.

- A schedule showing the earliest possible completion for each item of the Work and for the entire Work, all as indicated by the Milestone dates shown in Section 1.6. The Bidders should also provide manpower loading and cumulative weekly totals for the Work. Subcontractor man-hours shall be included and shown separately.

- A procedure for the handling and disposal of the Bidder's environmentally objectionable waste, both on and off site.

- The Bidder's proposed safety program for the Work.

- The Bidder's proposed quality program for the Work.

1.11   Execution of Contract

The undersigned agrees that this Tender is firm and subject to acceptance by MHS Engineering Services Inc. for a period of 30 days from the date of receipt of Tender and that if notified of award of Contract, he will:

1. Execute a contract with MHS Engineering Services Inc. in accordance with the terms and conditions noted in the documents.

2. Before commencing work, furnish to MHS Engineering Services Inc. copies of insurance policies and evidence of good standing with the Workmen's Compensation Board as required by the General Conditions of the Contract.

Name of Company: 33

Signed By: 34

Title: 35

Date: 36

Guideline 42

{The following is the tender form backup material and does not have to be returned by the contractor.}

Section II

Instructions to Bidders

2.1   Tenders

_____37_____ copy(ies) of drawings and _____38_____ copy(ies) of the specification are issued to each Bidder. Drawings and specifications are the property of MHS Engineering Services Inc. and shall be returned in good condition.

Tenders shall be delivered on or before _____39_____ p.m. Eastern Standard Time on _____40_____ to:

MHS Engineering Services Inc.   Marked: Tender Document
P.O. Box 8000   Inquiry Number: _____42_____
Anywhere, B.C.
L5R 3T9

Attention: _____41_____

Tenders shall be for a stipulated sum for all work as shown on the drawings and specifications, including all work to be done by subtrades, if any, required for the proper installation and completion of the Work in every respect. The Lump Sum Tender price shall be based upon rates expected during the scheduled construction period.

2.2   Examination of Tender Documents

Should the Bidder find discrepancies in or omissions from the drawings, specifications, or other documents, or should the intent or meaning of these documents appear unclear or ambiguous, the Bidder should at once notify MHS Engineering Services Inc. in writing and request clarification.

Guideline 43

{With the above statement you have to be very clear in your dealings with the bidders. If a Bidder has a better way of doing what you want done and has asked you for approval to do it his way, you do not issue an addendum. If it is a question that needs clarification about an error in the drawings, then an addendum is issued. In the first instance, it means the contractor has put some thought into what is required and it could mean savings for you. You do not penalize the contractor for being ingenious.}

Replies to such inquiries will be made in the form of a published addendum, which will be issued before closing time. These addenda will form part of the contract documents upon which this Tender is based.

MHS Engineering Services Inc. will not be responsible for any verbal instructions issued.

2.3   Site Inspection and Conditions

In addition to examining the Tender Documents, bidders shall carefully examine the Site of the proposed Work and make whatever arrangements are necessary to become fully informed of all existing and expected Site conditions. Bidders should also consider matters which, during the Contract time period, could affect the Work or performance of the Work in any way, especially anything affecting the cost of performing the Work.

For the purposes of examining the Site, a meeting will be held for all bidders on _____43_____ the _____44_____ day of _____45_____ at _____46_____, when the Owners will answer any questions by the bidders about the Work and the Contract Documents.

As a condition of the Contract, it shall be understood that the successful Bidder has visited the Site of the Work and satisfied himself as to the nature of the work and has included in his Tender a sum to cover the cost of all Work of the Contract.

Guideline 44

(You should be very leery of any bid from a contractor who has not visited the site, especially if he is not familiar with the plant. Sometimes, a contractor who is already working on site will do his own site tour and will be conversant in what you want, which is okay. You may still want to question him about the site conditions just to make sure he understands what the potential problems could be. Make a note that he did not take an official site tour with you at the bid meeting.)

The Tender shall be prepared with the understanding that work can be performed in accordance with all local labor agreements. Should the Contractor wish to work premium hours, he may do so at his own expense provided he has received prior approval from MHS Engineering Services Inc. personnel.

Personnel visiting this plant are required to wear safety headgear, safety glasses, and safety footwear.

Guideline 45

(You can add additional safety equipment requirements if you need to.)

No photographs shall be taken at the Plant without the permission of the Owner.

2.4   Award of Contract

MHS Engineering Services Inc. reserves the right, without liability, to reject any or all Tenders, to waive any informalities, and to accept any Tender which in their opinion is most satisfactory. MHS

Engineering Services Inc. reserves the right to reject Tenders received from parties who cannot show reasonable acquaintance with and preparation for the proper performance of the class of work specified herein. The Bidder must furnish evidence of such competency upon request.

Guideline 46

{You do not have to accept the lowest bid; however, you may have to explain to management why you did not. As stated previously, be careful whom you put on your bid list. If you do not want a bid from someone, don't give them the opportunity to bid. If you have any doubts about a contractor, check his references.}

It shall be understood by all Bidders that the Tender shall be valid, firm, and subject to acceptance by MHS Engineering Services Inc. and that no adjustment shall be made to the tendered amount for a period up to and including 30 calendar days from the stated closing time.

2.5   Taxes, Permits, and Fees

Each Bidder shall include in his Tender all federal sales taxes and all other customs, duties, and excise taxes levied on materials and equipment to be supplied under this Contract. The building permit is to be provided by Owner. The cost of all inspection fees and permits is to be included by the Contractor.

2.6   Pre-Award Meeting

After submission of Tender but prior to award of Contract, Bidders shall be prepared to review the Scope of Work with MHS Engineering Services Inc. Thereafter, no change shall be allowed unless approved, in writing, by MHS Engineering Services Inc.

Guideline 47

{This is a meeting with the successful bidder, or bidders if they are close in price. If there are several bidders, there are separate meetings and you can negotiate with the bidders to get the best price. This meeting is also your last chance to get work items included in the bid price. This is a common way of handling additional items you are aware of. If possible, send these additional items out to the bidders several days before a meeting. Be prepared for an extension of time in the bid closing date. You can always say no.}

2.7   Pre-Bid Contacts

Questions arising during bidding should be directed as follows:

1. For Site information and to arrange for visit, design, and Tender information:

    Mr./Ms.   **47**
    MHS Engineering Services Inc.
    P.O. Box 8000
    Anywhere, B.C.
    L5R 3T9

    Phone: 1 – 454 – 555 – 1212
    Fax: 1 – 454 – 555 – 1213

2. Terms and Conditions.

    Mr./Ms.   **48**
    MHS Engineering Services Inc.
    P.O. Box 8000
    Anywhere, B.C.
    L5R 3T9

    Phone: 1 – 454 – 555 – 1216
    Fax: 1 – 454 – 555 – 1217

2.8   Temporary Services

Office and Lunchroom

49

Washrooms

50

Parking

51

Power

52

Water

53

Lighting

54

Chapter

6

SCOPE OF WORK, DRAWINGS, AND SPECIFICATIONS

All projects revolve around the management of scope, schedule, and budget. Chapter 5 laid out the information needed from the contractor for the management of schedule and budget. This chapter guides you through the information that you will have to provide to the contractor to create and manage the scope. This scope of work was developed for a greenfield plant where the owner was providing the utilities for the contractors. The plant was constructed in northern British Columbia and at the construction peak there was a work force of approximately 570 people.

The greenfield plant was on the edge of a small town and the building lot was a forest when the project started. Following is the scope of the services and utilities that the owner had to provide before construction started:

- A contractors parking lot had to be built and it had to be expandable. Plugins for the cars in winter were not provided even though −40 C temperatures were anticipated.

- Roads around the site had to be built so that machinery could get around any road blockage.

- The owner was going to supply the sanitary facilities so a new potable water line was run onto the site. This was a tie-in to the firewater line. This provided water to the washcar and office facilities.

- The owner was also providing construction power so a power line had to be run in to the new construction power transformer. Power cable then had to be run to various locations around the site for the contractor's power panels.

- Telephone lines had to be run onto the site and a central junction box installed. This allowed the contractors and new plant to have phone service.

- A washcar with sanitary facilities for both men and women was rented and installed in a central location. (You do not want people wasting time walking too far.) There was no sanitary sewer close to the site so septic tanks were used.

- As it was a clear site and would eventually have a large warehouse, the surveyor ran in three different survey points so there would be reference points that could be seen as the building was erected. Once the building got to a point where the reference points could not be seen, the surveyor came back and moved the points inside the building.

- A construction office had to be built using rental trailers. They were set up so that more trailers could be added should the need arise.

- As the owner was the general contractor they were responsible for the first aid room for the whole site (570 people). This first aid room was built inside the main office trailer. This was staffed by a safety officer hired by the owner.

- Areas for laydown of materials and contractors trailers had to be leveled and gravel put down.

- A security perimeter had to be fenced and a security gate set up. Security had to be arranged for nights and weekends. We visited the local fire department to let them know we were out there and had them come out for a visit to see what the site looked like.

- Temporary outdoor pole lighting had to be installed.

- Signage was made to direct traffic and visitors as well as posting the land with No Trespassing signs.

- Contracts had to be let for emptying the septic tanks, snow clearing, and cleaning the offices and washcar.

- Two way radios were used so a permit had to be applied for.

In an operating plant all of the above is in place. When you hire a contractor part of the scope of work is what services the plant is going to supply and what ones the contractor will have to supply. This scope of utilities and services is usually glossed over; however, they are part of the scope of work and this chapter will guide you through the process of developing a thorough scope of work, including utilities and services.

Guideline 48

{No matter which type of contract is used it is imperative that the Scope of Work be well defined; that all drawings, specifications, and standards are listed; and that any specific plant requirements be spelled out, e.g., safety, access, use of facilities, etc. This is extremely important and cannot be emphasized enough, as most projects run into trouble because someone didn't handle this section properly. If you have some unresolved design issues when preparing the scope, include something and change it later. Do not be ambiguous.}

Scope 1   General Description

Without limiting the "Definition of the Work" as set forth in Scope 2, the work of the Contract consists in general of performing all the tasks required to complete the items listed on the attached sheets and as indicated on the Construction Drawings and as described in the Contract Documents.

Scope 2   Definition of the Work

Except as provided for in Scope 4 "Not Included in the Work," the work shall consist of supplying all other materials, construction equipment, licenses, tools, temporary facilities, weather protection, temporary enclosures, temporary heating and lighting, access within work areas, and providing supervision, labor, overheads, permits, (except building permits and environmental permits), travel, accommodations, and everything necessary to accomplish the work called for in the Contract documents.

Guideline 49

{The intent of the above two statements is to prevent the contractor from trying to add extra charges for items not specifically detailed in the drawings—items he should know are needed to complete the work. For example, you do not show formwork on a concrete drawing but it is needed to pour the concrete.}

Section III

Scope 3   Included in the Work

3.1   The following list of work to be performed by the Contractor is presented to complement or clarify the Construction Drawings, Specifications, and other Contract Documents, but shall not limit the "Definition of the Work" as described in Scope 2, nor shall it constitute a complete list of the work.

3.1.1   General

1.   Providing the Owner with proof of insurance prior to commencing work.

Guideline 50

{Your plant requires this to ensure that the plant property is protected. Make sure you get this before the contractor starts work and if it expires during the contract period, make sure you get a copy of the new policy. Do not let a contractor work on your site without proper insurance coverage.}

2.   Prior to commencing work and prior to receiving final payment, provide the Owner with evidence of compliance with the requirements of the workers compensation insurance including payments due thereunder.

(a)   At any time during the term of the Contract, when requested by the Owner, the Contractor shall provide such evidence for himself and his Subcontractors.

Guideline 51

{You need the original letter from the Compensation Board, not a copy. People have been known to change dates on copies and the contractor may be in arrears. In certain jurisdictions in Canada, if the contractor is in arrears and the

Compensation Board finds out he is working for you, you are automatically responsible for his back payments. You can ask for a proof of payment with every invoice if you so desire. Everyone on your site should be covered by workers compensation insurance. If you are using an individual consultant, he may not be able to get coverage because he is a one-person operation. In these cases you can add them to the plant coverage by advising the Compensation Board in writing. All injuries should be reported, as even seemingly minor injuries can sometimes have long term affects on a person. Making a claim at the time of the accident could help prove a compensation claim years later for a related illness.}

3. Performing general work items listed in the General Conditions and Special Conditions.

4. Accepting the Site in its existing condition at the time work commences.

5. Submitting to the Owner, prior to commencing the Work on the Site, a list of the names of known supervisory, administrative, and other employees required on the Site and reporting any changes to the Owner as they occur during the duration of the Contract.

Guideline 52

{This information is required for plant security purposes. In case of emergency, you want to know who is on site and where they are located on the site. Also, there may be people you do not want on site and you can tell the contractor to keep them away. It is better to know *before* they show up on the site.}

6. Maintaining complete records at all times during the progress of the Work, in a form and detail of presentation acceptable to the Owner.

7. Submitting to the Owner a list of temporary buildings and building sizes that the Contractor intends to bring onto the Site, before mobilization of the structures. The structures should be propane heated. Owner will provide and hook up 120 volt power for lighting purposes only. The Owner will advise the Contractor of the Site where the temporary buildings will be located.

Guideline 53

{On most sites space is at a premium; therefore, you want to know in advance what the contractors are bringing on site. You will have to assign a location for him along with the other contractors you have on site. You will control the location of all trailers, washcars, and storage containers. Normally, you try to locate the trailers close to the work site. When assigning a site, try to envision what the site will be like when the trailer has to be removed at the end of the job. Will it be boxed in? Will it be in the way of landscaping? You may have to move the trailers before the end of the job if they are in the way.}

8. Providing First Aid personnel, equipment, and supplies, pursuant to Special Conditions and complying with the Province/State of ___55___ Occupational Health and Safety Act First-Aid Regulations for the duration of the Contract.

Guideline 54

{The first aid requirements increase with the number of people on site and as the distance from a hospital increases. If you are using the plant first aid facilities, you can charge the cost back to the contractors prorated by the number of people they have. You will, however, get the complaint that "my crews are safer than theirs, therefore we should not have to pay as much." This has to be an across-the-board decision, and it has to be sorted out before you start.

If you have 300 people on site, the general contractor should have first aid facilities for 300. The subcontractors should have first aid facilities for the number of people they have. So it would not be uncommon to have several first aid rooms around the site. Each contractor should have the proper class of first aid attendant required. Every jurisdiction is different and first aid tickets in one area are not necessarily valid in another area. You should understand what is required and make sure the requirements are fulfilled throughout the project. Do not count on the contractors to understand the requirements, as the legal wording can sometimes be confusing and not completely understood by some of them. If there is ever an accident, you want to make sure you have done your part correctly. You or your designate should look at the first aid journals to make sure all accidents or injuries are being reported. Make sure you have all this sorted out before awarding any contracts.}

9. Supplying and installing equipment, materials, and labor to connect and distribute construction and potable water required by the Contractor for the work.

Guideline 55

{Leave it up to the contractor to provide everything necessary to get water to his work site. At the site meeting show the contractors where they can obtain the required water. Do not let contractors use plant fire hoses to transfer water, as the hoses are easily damaged and will not be available in case of a fire. Generally it is not a problem if the contractor uses plant fire hydrants; however, check your plant's policy on this as there may be flow indicators that set of the alarms when a fire hydrant is used. If their use is allowed, test the operation in the presence of the contractor before he uses it. They are easily broken when not operated properly. Do not let the contractor use any of the plant hydrant fittings or wrenches, as you may need them and you want them to be in their proper place. In some plants potable water and fire water are the same i.e. the potable water line with a backflow preventor is tied in to the fire line}

10. Providing any wash trailer, lunch trailers, and sanitary facilities required by the Contractor for the Work in addition to those provided by the Owner.

Guideline 56

{Most plants are reluctant to allow contractors to use their employees' facilities. You should not even consider it except under very unusual circumstances. If a contractor is doing fiberglass work, clean and dirty areas, washers and dryers, and showers are needed. Plants are not equipped for this.}

11. Providing all gas or oil heating products required by the Contractor for the Work.

Guideline 57

{Do not provide gas, diesel, propane, or fuel oil for heating. Let the contractor look after himself.}

12. Providing construction power panels as required for the Work. The Contractor should include power requirements with bid. Where Contractors cables have to be attached to Owners power supply, Owner's personnel will do the tie-in.

Guideline 58

{A construction power panel is a panel the contractor makes to provide his power requirements. It will have a transformer mounted on it to reduce the voltage down to 480/240 V and 120V. There will be several 120 V plugs to run hand tools and temporary lighting and it could also have a welding receptacle on it. The contractor moves this panel around the site as required. The power to it can come from different locations depending how the plant is set up. Sometimes the panel will have a welding plug on it for directly plugging into one of the plant's welding receptacles, or it can have pigtails that allow it to be tied into a splitter box or a MCC.}

13. Providing electric, gas, or diesel welding machines required for the Work. Owner will determine type of unit.

Guideline 59

{If you planned work during a shutdown, the plant maintenance may be using the welding outlet the contractor was planning on using. This could result in paying extra for a different type of welder. It is best to have the contractor use gas or diesel powered welders. Make sure he has long enough leads so that the units can be outside when operating.}

14. Providing compressed air for construction purposes, as required for the Work.

Guideline 60

{The contractor should not be tied into the plant air system. Most plant air systems are running at capacity and any additional load can have an affect on the plant operations, usually at the most inopportune time. If the contractor is setting up a calibration shop check to see if the plant air can handle the additional, although small, load.}

15. Providing telephone and other communication facilities the Contractor may require for the Work.

Guideline 61

{With the advent of the cell phone, the telephone use problem can be solved, but there is still the problem of the fax. Generally, you do not want the contractor using your fax machines. Other than cost, there is the matter of your privacy as well as his. Faxes are always lying around fax machines waiting for pickup, and you do not want outsiders reading them. Vendors on site for a short time may want to use your fax. In this case make arrangements for your secretary to send them or make sure there is nothing lying around.}

16. Providing temporary lighting as required for the Work.

Guideline 62

{Do not get involved in providing temporary lighting for any reason. The only lighting you provide is what is there now. If you have temporary lighting for large repeatable maintenance tasks that you will allow the contractor to use, make sure the contractor understands exactly what you will provide.}

17. Providing covered shelter, where necessary, for storage of the Contractor's own supplied equipment and material.

Guideline 63

{The contractor should provide his own tool shed. Common sheds are highway trailers or containers. They are usually placed close to the job site so the workers don't have to go too far to get what they want. They may not be with the other trailers; in fact, they may end up in the plant if space is available.}

18. Providing and using facilities including hoarding and like structures as required to protect the Work and to enable the Work to proceed during inclement weather. This would include enclosures and temporary heat to protect and thaw any frozen ground and for curing of concrete during the Work.

Guideline 64

{Look closely at this clause. If you will be working during cold periods, the amount of risk the contractor builds in to cover heating and hoarding may be too expensive. Under these circumstances, it is best to handle heating and hoarding on a time and materials basis. Keep in mind that this is a very expensive exercise. For rain, the contractor should have some risk money built into his price.}

19. Providing dewatering equipment and performing dewatering operations as required to facilitate the Work.

Guideline 65

{As with the previous item, if you suspect groundwater in your excavations, you may want to handle this on a time and material basis. In plants, you can have spills that fill excavations, your floor trenches leak, or you may be beside a river and have a high groundwater problem.}

20. Constructing and maintaining barriers, guard rails, covers, and lighting as required for the Work, in conformance with the safety regulations or as directed by the Owner.

Guideline 66

{The contractor has to keep the site safe to the plant's satisfaction. If the contractor does not keep the site safe and is in violation of regulations, shut the job down and have your safety group talk to him. If you do not have a safety group, get your local workers compensation board representative involved. As a last resort have the contractor removed from the site. Be proactive in this issue.}

21. Maintaining Site cleanliness pursuant to General Conditions and the Owner's Site environmental policy.

Guideline 67

{A big problem is packing crates. They are usually good wood that everyone wants. If you start giving the wood away it gets out of hand and [other?] good material soon starts disappearing. You will have a steady stream of people looking for material. Your best option is to burn the material on site. If burning is not allowed, then come up with a secure method of getting the material to the local landfill or use the plant's landfill. Tell the contractor up front what you plan on doing with the material. Do not give anything away.}

22. Providing the Owner proof, if requested before use, that materials supplied by the Contractor are the types and quality called for in the Specifications.

Guideline 68

{This is a quality check on your part. There are counterfeit items, poor quality items, and people who do not understand the specifications. If you are suspicious about anything, check.}

23. Maintaining a sufficient stock of materials on the Site at all times to meet the demands of the Schedule with a reasonable reserve to compensate for changes in the Work or changes in the construction program.

Guideline 69

{You do not want contractors getting materials from the plant stores. This is difficult to control and your charge number can get into the wrong hands. It is okay for you to supply the odd item, but as a general rule don't let the contractor into your plant stores. For projects, you are better off to be self-contained and have your own arrangements with suppliers.}

24. Transporting to the Site of the Work, rehandling from storage areas on Site, and transporting to the point of use and offloading all equipment and material required for the Work. Storing and locating equipment and materials in areas designated by the Owner.

Guideline 70

{The contractor should off load your material in the storage area and eventually move it to the site. When you are out for bid you should have a storage area picked out. If you live in an area that has snow, the contractor should provide dunnage to place pipe on so the pipe is out of the snow and you can find the pipe when you need it.}

25. Notifying the Owner and the local authority having jurisdiction sufficiently in advance when any testing is to be done or any inspection is required. Assisting the Owner and the local authority, as required, with any testing or inspection they wish to carry out.

Guideline 71

{The contractor should look after certain on site testing, typically pipe pressure testing and electrical installation. If pressure testing pipe, then the inspector has to be called and a time arranged. At the appropriate time the pipe is filled with the test medium, pressure applied, and the pipe inspected.}

26. Establishing lines and grades required to set out the Work from the basic reference lines and benchmarks established by the Owner.

Guideline 72

{Have your own surveyor run in the basic reference lines and elevations. You may have to run in several locations in order to cover the total site. Your surveyor should be used to backcheck important pieces of work to verify the contractor's surveying.}

27. Checking out work performed by others as it affects the Work, bringing any discrepancies, errors, or deficiencies found in such work to the immediate attention of the Owner, in writing, prior to commencing that portion of the Work affected.

Guideline 73

{If the contractor is going to place a piece of equipment on a foundation built by another contractor, he should check the elevations to make sure the foundation is where the drawing calls for it to be. If it is not, the contractor should get back to you right away so you can get the other contractor to fix the problem.}

28. Making good the Work, work done by others, structures and services the Contractor disturbs or otherwise affects during the performance of the Work. The affected Work, work done by others, structures and services shall be made good by the Contractor, in every trade, to match the adjacent work to the satisfaction of the Owner.

Guideline 74

{The contractor should repair any damage he or his subcontractors do to the plant or to work someone else has installed. The contractor's insurance policy will usually cover this item, so make sure the contractor has the insurance coverage your tender document requires.}

29. Coordinating with other contractors working in the same areas to allow access to the areas where they are required to work and to prevent interference with or interruption of their work.

Guideline 75

{This is a persistent problem on job sites and requires careful coordination on your part to stay out of trouble. Every contractor feels he is the top contractor on the site and therefore he should get into the areas when he wants to. You have to have coordination meetings every week to control who is where when. You will have to referee disputes. Look at which delay will have the least affect on the schedule. Is there another way of doing it or can one contractor work overtime to do it or finish it faster?}

30. In accordance with the Site policy for handling and disposing of environmental objectionable waste, the Contractor shall:

   (a) dispose of off site Contractor's tires, batteries, oil, fuel, and toxic waste and any other environmentally objectionable waste materials.
   (b) dispose of on site paper products to be recycled, in a bin or container provided by Owner for pick-up by Owner.
   (c) separate lumber, concrete, and plastics in designated areas on site. Also use putrescible waste location as appropriate; and
   (d) excavate earth contaminated by oil spills or other contaminates and dispose of the contaminated materials off site.

Guideline 76

{You do not want the contractor's environmental garbage and the problems associated with it. Be vigilant about environmental garbage and keep it out of the waste stream. You will run into trouble with the landfill personnel if they find material they are not allowed to bury. If you have garbage pickup at your plant, bring in extra bins and make arrangements for additional pickup. Most plants use the city garbage system; however, some plants have their own landfills. You may want to sort your wastes between construction waste and food waste. This way you can control pests by having the food waste picked up on a more regular basis.}

31. Obtaining all necessary permits, business, service and operating licenses, and inspection certificates required for the Work.

Guideline 77

{The plant obtains only a few permits directly--a building permit (if needed) and any environmental permits. The contractor can assist in assembling these, but it is usually the plant that acquires them.}

32. Maintaining Site security for the Contractor-supplied equipment, tools, and material for the duration of the Contract period. Contractor is to provide the Owner with a list of equipment and tools, including tradesmen's tools, that will be brought on the Site. A copy of this list should be given to Security when Contractor comes on Site (see attached form).

Guideline 78

{You do not guarantee security for the contractor's equipment and material. Even if your plant has a security guard, that is no guarantee that the site is protected. The contractor has to take the necessary precautions to look after his own equipment and material. You may have to make arrangements for your plant security to check all the areas during off-hours, but this for the plant's protection, not the contractor's.}

33. Providing the Owner with a detailed set of Construction Schedules pursuant to General Conditions, prior to commencing Work on Site.

Guideline 79

{Let the contractor develop and maintain the project schedule, based on your milestone dates. Get this overall schedule before they start work to make sure they have an understanding of what you want and how they plan on approaching the project. For managing the project, they will usually prepare a two or three week look-ahead schedule which is reviewed at the weekly project meetings.}

34. Submitting a Quality Control Program, tailored to the Work, to the Owner for approval, prior to commencing Work on the Site. Subsequently performing the Work and all quality testing necessary in accordance with the approved Quality control program. Verifying that inspections and tests are performed. Carrying out in-process evaluation of the quality of construction. Submitting quality control program compliance, inspection, test and evaluation records to the Owner on a regular basis to ensure the Work is performed in accordance with the Construction Drawings, Specifications, and other Contract Documents. Notwithstanding anything to the contrary in the Specifications or elsewhere in the Contract Documents, the Contractor shall be responsible for all quality control testing and inspection. The Owner will perform quality assurance (QA) auditing to verify the Contractor's conformance to the approved quality control program.

Guideline 80

{If the contractor is installing pipe that is designed to ASME B31.1 Power Piping Code, he should provide you with all the written procedures, forms, testing program, and inspection program he will be using to ensure the piping is installed according to the code. Since all construction is covered by some type of code, standard, or regulation, there should be a quality control program to cover all construction items you will encounter. Small contractors may not have one and you may have to assist them in developing one. It is prudent to have a quality control program from all contractors.}

Testing by the Contractor shall comply with the following:

i.   Testing required to provide quality control to ensure the Work strictly complies with the Contract Documents shall include, but not be limited to:
   (a) testing for alternative products;
   (b) all testing specified in the Contract Documents; and
   (c) any other testing required as a condition for deviation from the specified Contract procedures.

ii. The Contractor shall be fully responsible and bear all costs for all quality testing and shall conduct such testing in the following manner:
   (a) Providing testing facilities and personnel for the tests and informing the Owner in advance to enable the Owner to attend;
   (b) Notifying the Owner when sampling will be conducted;
   (c) Within five days of completion of testing, submitting test results to the Owner with a written interpretation of the results; and
   (d) Identifying test reports with the name and address of the organization performing all tests and the date of the tests.

iii. Approval of tested samples shall be for the characteristics named in such approval and shall not change or modify any Contract requirements.

Guideline 81

{Example: If you are approving steel mill test reports for steel composition, then you are not approving any other characteristics that may be shown on the test report, such as hardness, etc.}

iv. Contractor's testing agencies, their inspectors, or representatives are not authorized to revoke, relax, enlarge, or release any requirement of the Contract Documents nor to approve or accept any part of the Work.

35. Transporting to and from the Site of the Work all Contractor's and Subcontractor's employees and other persons the Contractor requires during the performance of the Work.

Guideline 82

{If you talk to the contractor you may find that his employees cannot be required to walk more than a quarter of a mile from their car to their work location or trailer. If you are in a large plant, all the workers will probably have to be shuttled from the parking lot to the construction trailers. This is the contractor's responsibility; however, you have to advise him before contract award where his trailers will be located and where his employees will park. If you want to limit the number of vehicles on site, the shuttle vehicle should be a working vehicle and not one reserved solely for moving the workers to and from the front gate.}

36. Coordinating the activities of the Contractor with the Owner's personnel.

Guideline 83

{It is your responsibility to deal with the plant personnel when it comes to the contractor's activities and coordination with the plant.}

3.1.2   Work to be Done

Guideline 84

{This section is a description of the actual work you want the contractor to do.}

Scope 4   Not Included in the Work

4.1   The Owner will provide the Contractor with the following items, free of charge at the Site:

> **Guideline 85**
>
> {The plant does not have to provide any of the following items. It should be on a project-by-project basis. These are all services the plant already has in place, and for small projects it is cost effective to provide the contractor with some of the services. For large projects it is advantageous for the contractor to look after himself.}

a)   Construction power

> **Guideline 86**
>
> {The easiest way to handle this is for you to look after and pay for it. If you try to backcharge the contractors, it can become a nightmare and is not worth the effort. The contractors should provide their own construction power panels. You should supply the panel connection point only and the contractor can supply all the cable necessary to get the power to his panel. It helps if you have more than one connection location so the cable lengths are kept to a minimum and there are enough connections for all contractors.}

b)   Construction and potable water

> **Guideline 87**
>
> {You are only providing a tie-in point for the contractor to hook his hoses too. You are not providing hoses. The contractor should take care of any bottled water he needs. If possible, find a potable water location that keeps the contractor outside of the plant.}

c) Construction material storage area

Guideline 88

{This usually refers to outside storage, not inside storage or secure lockup. Be very explicit with the contractor as to what you are providing. Inside storage may involve taking up some of your plant stores area, which could be difficult.}

d) Sanitation at locations advised

Guideline 89

{The contractor should provide his own washrooms unless the plant is willing to allow the contractor into the plant. This is never a good idea as damage and cleanliness will be blamed on the contractor and you will probably end up banning the contractor from the plant washrooms. The only exception is when you need female facilities. You can usually get away with using the plant washrooms for women as there are very few women with the contractor and the added expense of additional contractor facilities is not worth it.}

e) Owner's first-aid pursuant to Special Conditions

Guideline 90

{The contractor has to provide first aid according to the plant regulations. The requirements are usually based on the number of people the contractor has and how far the job site is from a hospital. The plant first aid should be used as a last resort; otherwise, you may be saddled with first aid costs for materials used to treat the contractor's first aid cases. There will be the occasional use of plant first aid as the plant attendant may be the closest one available, but it should not become a habit.}

f) Garbage boxes located outside the buildings for the Contractor to deposit his construction waste, excluding environmentally objectionable waste

Guideline 91

{Make sure the contractor is not disposing of heavy waste in dumpster boxes. He should provide his own. Personnel cleaning up will throw as much as they can in the boxes, and if the box is too heavy the truck may not be able to pick it up. Be careful as the contractor may be doing machinery repairs on site and could leave you with his left over garbage such as oil, oil filters, grease cartridges, etc.}

g) Insulation removal and reinstallation

Guideline 92

{Insulation is a specialty trade. Hire an insulation contractor to go around and expose those areas of the piping that require work by other contractors and to return when the pipe work is finished to reinsulate it. You will get a much better job this way.}

h) A crane contractor whom the Contractors will be required to use

Guideline 93

{If you are in a semi-remote or remote location, it may be less expensive to make a deal with a crane contractor to provide all the necessary cranes rather than have each contractor bring in his own crane contractor. The crane costs are backcharged to the contractor.}

i) A scaffolding contractor whom the Contractors will be required to use

Guideline 94

{As with cranes, it may be more economical to bring in a scaffolding contractor to handle all the scaffolding needs rather than have each contractor supply his own.}

j) Tie-in to Owner's power supply

Guideline 95

{For safety and operations reasons you do not want contractors opening plant electrical panels or MCCs and wiring their leads into your plant power system. Most mechanical contractors do not have the qualified personnel to do this. Let your plant electricians do this part of the job.}

Section IV

Contract Drawings

Guideline 96

{This section is for listing the drawings that apply to the work. They can include any drawing or sketch, old or new, that you feel helps the contractor better understand the extent of the work.}

1.1   General

1.1.1   The drawings listed below are the Contract Drawings and together with the other Contract Documents described the Work as defined in Scope 2 "Definition of the Work."

1.1.2   The drawings and specifications referred to herein offer no guarantee or assurance of indicating or describing every detail of the Work required. The Contractor shall make allowances in determining the value of those things not necessarily described nor mentioned but they find to be essential to the Work.

Guideline 97

{This wording is used to preclude extra charges for items understood to be required but not specified because standard engineering/construction practice would include it. An example would be welding rods. Construction practice is to keep them in an oven before use. The use of the oven is not called up on the drawings nor in the specifications. A contractor will test you on this so be prepared.}

1.2   **Drawings**

<u>Drawing Number</u>   <u>Issue</u>   <u>Rev</u>   <u>Title</u>

Guideline 98

{In this section, list all relevant drawings and sketches. Each one should have a drawing number that is distinct from the others. Issue or revision numbers should be used and indicated, and the titles should be written out in full.}

Section V

Specifications and Standards

Guideline 99

{This section is for listing all the specifications and standards you feel apply to the work. These are usually drawings and procedures that have been standardized so you get consistent outcome in certain parts of the work. You can use standards developed for the original plant construction or from other companies. They should be modified for your particular application.}

Number   Issue   Rev   Title

Guideline 100

{As with drawings, the specification and standard references used should be distinct from the others, should have an issue or revision number, and the titles should be written out in full. Sometimes you will have standards without issue or revision numbers on them. In this case, use the date of original issue as the identifier.}

Chapter

7

AGREEMENT AND GENERAL CONDITIONS OF CONTRACT

Section VI

Guideline 101

{This section is used to specify what contract document will control the project. This book references Canadian Construction Document Committee (CCDC) contracts (see Chapter 4); however, if your plant has its own contract, reference that. Make sure the contractors bidding on the work are familiar with your contract. If not give them a copy of it before you award them a contract to do the work.

The CCDC contract contains the following articles:

> Articles of Agreement
> General Conditions
>> General Provisions
>> Administration of the Contract
>> Execution of the Work
>> Allowances
>> Payment
>> Changes to the Work
>> Default Notice
>> Dispute Resolution
>> Protection of Persons and Property

Governing Regulations
Insurance and Bonds
Indemnification-Waiver-Warranty

These articles are not covered in the Tender Document. You will have to change your Tender Document to make it compatible with the contract you intend to use. Following are some changes that can be made to the CCDC2 contract document to reflect the plant conditions:

6.1   The terms and conditions as detailed in the Standard Construction Document CCDC No. 2 (Stipulated Price Contract), latest edition, including Supplementary General Conditions for use in The Province of British Columbia, shall apply to this contract (document not supplied).

Guideline 102

{This is where you reference the contract you want to use.}

6.2   In Standard Construction Document CCDC No. 2 the term Consultant may not apply and should be deleted on a project by project basis.

6.3   Standard Clause GC. 20 Insurance, of Standard Construction Document CCDC No. 2 shall be considered deleted and the revised section GC. 20 enclosed with this inquiry shall apply.

Guideline 103

{If you are using a plant contract, this clause is not required. If using a different contract, the insurance and bonds section should reflect what the plant requires.}

GC. 20   **Insurance & Bonds**

20.1   The Contractor will, if requested by the Owner, provide the Owner with either or both a Performance Bond or Labor and Materials Bond in respect of the contractor's obligations under the Contract. Such bonds shall be in the form and in such amounts as the Owner may reasonably require from surety companies authorized to carry on the surety business in British Columbia.

Guideline 104

{A bond is a contract between the bonding company, the contractor, and the plant; it is not an insurance policy. The bonding company agrees to guarantee that the contractor will perform as specified. The performance bond is a guarantee that the contractor will perform his contract obligations. A labor and material bond is a guarantee that the contractor's suppliers and labor will be paid.}

20.2   The Contractor shall maintain during the performance of the Work, in form and with carriers acceptable to the Owner:
(a) Comprehensive General Liability Insurance, and
(b) Insurance covering all motor vehicles or other craft utilized by the Contractor, and
(c) Insurance covering its equipment used in connection with the Work.

Guideline 105

{Contractors' employees bringing their personal vehicles on the plant property must have liability insurance that meets the plant requirements. The contractor requires vehicle insurance, per the insurance document and the employee bringing his vehicle on site also requires the same vehicle insurance per company policy. If the employee's vehicle does not have the required level of insurance, the vehicle is not allowed on site. Insurance should be checked as you can run into trouble if the employee damages plant property and disrupts plant operations.}

The limits of liability insurance described in (a) and (b) shall be $2,000,000 per occurrence. The insurance referred to in (a) shall name the Owner, its directors, officers, and employees as named insureds, and shall include liability and completed operations coverage. The insurance referred to in (b) and (c) shall waive rights of subrogation against the Owner, its directors, officers, and employees.

Guideline 106

{The liability amount of $2,000,000 should be checked as it may be too low.}

20.3   The Contractor shall provide the Owner with certificates evidencing all required insurance prior to starting the Work. All insurance shall provide that the Owner receives 15 days prior notice of any change or cancellation.

Guideline 107

{Check your company procedure regarding the requirement of bonds. Also check the limits and types of the insurance required. Make sure you have copies of these documents before the contractor commences work.}

Special Conditions for Site Contractors

1. Job Conditions

a) The Contractor agrees that he is familiar with the premises, drawings, and specifications, accepts the conditions that will exist in performing the Work, and the price for this Contract was established with full consideration of such acceptance. The Contractor shall perform the work under the direction and to the satisfaction of the "Owner's representative" designated in a written notice from Owner to Contractor. The Contractor shall cooperate with the Owner's representative and other contractors on the premises and shall carry on the work as not to hinder, delay, or interfere with the operations of the Plant or other Contractors.

Guideline 108

{Advise the contractor who has the authority to act on the plant's behalf in the normal management of the project. You may even outline that person's responsibility, e.g., can he issue field work orders? This can be done at a meeting and recorded in the minutes or you can advise in writing. There has to be <u>one</u> person in charge from the plant to prevent confusion.}

b) The Contractor shall bring to the immediate attention of Owner's representative any discrepancies or errors found in the work performed by others.

Guideline 109

{The contractor should make sure, before he starts his work, when interfacing with someone else's work that the work he is interfacing with is correct.

Examples would be:

• A mechanical contractor placing a tank on a foundation done in error by a civil contractor.

• An electrical contractor attempting to hook up a motor supplied by others, with the connection box on the wrong side of the motor.}

c)   Contractor shall secure and pay for all permits (excluding environmental permits and building permits) and licenses necessary to accomplish the work.

Guideline 110

{Get copies of all permits for your files. This is easily overlooked. Get a list of permits the contractor will get to ensure that all permits are applied for. The known permit fees would be included in the quote; however, there may be occasions when, as construction develops, an additional permit may be required. In this case you should pay for the extra permit. An example would be having to get an additional highway crossing or wide load permit to get material to the job site using a method that was not contemplated at the tendering stage.}

d)   Except as otherwise stipulated in the contract documents, Contractor shall provide and pay for all inspection fees necessary for the performance of the Project and shall give all inspection certificates to Owner's representative at the site.

Guideline 111

{The types of permits are usually known by the contractor; however, if you are aware of special permits that the contractor would not know about, you should pay for them and advise the contractor accordingly.}

e)   Contractor shall test the work of the Project as specified and shall submit test certificates to Owner's representative for each test stating the test results as compared to the specified results, the date of the test, and the name of the person conducting the test.

Guideline 112

{If the test is critical, the plant may want to do the testing, in which case the contractor should be advised accordingly. An example would be a new fire main. The contractor may do a test so he knows it will meet the standards. Then the plant can bring in their own inspector to perform the test.}

f)   Contractor shall supply all materials and equipment required for testing including equipment for filling, heating, draining, discharging, or otherwise required for testing work of the project, and shall provide all water, air, steam, or other material required for the test unless otherwise indicated.

Guideline 113

{The contractor may come to you for some of these test liquids/gases. Make sure the plant has the capacity to provide them. Sometimes it can take days of constant flow to fill a vessel for testing. You do not want to hinder the plant operation for this, but sometimes you have no choice.}

g)   Contractor shall remove snow from the work areas and from Contractor's storage areas on site, as required for the work of the Project.

Guideline 114

{Do not get involved with clearing snow for the contractors. If the plant has a contractor for snow removal, do not offer his services, and keep snow clearing to what the plant usually does. If the contractor wants to make a side deal with the plant contractor, that is fine, but do not get involved unless it affects the plant.}

2. Supervision

a) Contractor shall keep on the project continuously during the progress of the Work a competent general superintendent and any necessary assistants, all to be satisfactory to the Owner's representative. The Superintendent shall not be removed from the project and another substituted for him except with the consent of the Owner's representative unless the Superintendent ceases to be in the Contractor's employ. The Superintendent shall represent Contractor in his absence and all directions given to the Superintendent shall be as binding as if given to the Contractor.

Guideline 115

{You do not want the contractor to take a good supervisor off your job and send him to another project. Keep on top of this, as contractors will hide a superintendent on a job site while waiting for his real assignment to start up on another project. If the superintendent is good, you do not want him removed to the detriment of your project.}

3. Temporary Services and Facilities

a) Unless otherwise stipulated, the Contractor may use the following services at locations on the site as determined by the Owner's representative. All hoses or cables, etc. required to run from the service location to work site will be supplied by the Contractor.

i. Electrical power is 120V, single phase, AC and 600V, three phase, 60Hz. Contractors shall be notified by Owner's representative of any power outage, times, and dates. Contractor to provide their own construction power panels which will be hooked up by the Owner. Contractor to advise Owner of power requirements prior to contract award.

Guideline 116

{This is the power set-up used in Canada. Change this stipulation to suit your country or plant. In order to minimize construction disruptions, keep power outages to weekends or overnight.}

ii) Electrical power for welding machines subject to the type of units required, as determined by the Owner's representative and the Contractor. Contractor may be required to use gas or diesel welding machines.

Guideline 117

{It is easier all around to have the contractor provide gas or diesel welding machines. Using a plant wall outlet can lead to problems with the plant maintenance crews. They usually want the outlet just as the contractor plugs his lead in. Since the plant has control of the outlets, the contractor will be expected to move his cable. Therefore, start out right and have the contractor provide a gas or diesel welding machine.}

iii) Water for construction at various locations on the site.

Guideline 118

{If the project is a greenfield plant, you will probably install water taps at various locations for the contractors' use.}

iv) Mill air for construction at various locations on site.

Guideline 119

{This would be a minimum amount of air, possibly for clean up with an air hose and a calibration shop. Contractors should be using electric tools, not pneumatic.

If the contractor requires large quantities of air, he should bring in a compressor.}

v) Sanitary facilities provided and maintained for the use of all Contractors on the site.

Guideline 120

{If the project is fairly large it may be convenient for the plant to bring in washcars for all the contractors to use. The plant would maintain and clean them for all the contractors. The cost would then be prorated back to the contractors based on the number of employees they have on site.}

vi) First aid room, ambulance, and emergency response team provided for the use of all Contractors on site

Guideline 121

{The plant first aid room should be used as a last resort. The cost of excessive visits should be backcharged to the contractor. Keep in mind that the contractor has included the cost of operating his own first aid station and if he uses the plant he is saving money at your expense.}

vii) Owner will only provide building lighting. Contractor to provide temporary lighting as required.

Guideline 122

{The only exception would be temporary floodlights that run off their own generator. They are usually rented and if the plant has arrangements for inexpensive rentals then go ahead and rent it; otherwise, do not supply temporary lights.}

b) Contractor will provide his own trailers for lunchroom, office, etc. The Owner will provide and hook up 120 V power for trailer lighting only. Trailers should be equipped with propane heaters.

Guideline 123

{Electric heaters in trailers use too much energy and it is recommended that they not be used. By using propane, the contractor is responsible for looking after the heating arrangements and he can handle the costs accordingly. Review the above items carefully and change them so that they match Section 2.8 Temporary Services in the Tender Document.}

4. Responsibility for Errors and Deviations

a) Contractor shall check and verify all field measurements and submit for approval all shop or setting drawings and schedules. Owner's representative's approval of such drawings and schedules shall not relieve Contractor from responsibilities for deviations from the plans and specifications unless Contractor has in writing called Owner's representative's attention to such deviations at the time of submission and has secured Owner's representative's written approval, nor shall it relieve Contractor from responsibility for errors in shop drawings or schedules.

Guideline 124

{The schedules referred to are not about time and sequence of operation but are lists of materials. An example would be a cable schedule, which is a list of all the cable used on the job. It would show enough information so that the cable could be purchased from the schedule. When you sign off or approve the drawings and schedules, the contractor is still responsible for meeting the plant standards. If the contractor puts in writing that he is deviating from the project plan and you sign it, then the contractor is not responsible for the results of that change.}

b) Upon completion of the Project, Contractor shall furnish Owner with three prints of each drawing marked to show the exact location of all things not shown in detail on the construction drawing. This might include items such as pipe, conduit and tubing runs, grounding wire, and locally mounted push buttons and instruments or items for which the location was changed on the written instruction of the Owner's representative but was not shown on a revised construction drawing. Upon request of Contractor, Owner's representative will furnish him free of charge with a reproducible print of such construction drawings for marking up as setting, erection, or as-built drawings.

Guideline 125

{Make sure you know where all underground pipes and cables are. Have the contractor make up what are called "as-built" drawings. These are drawings that reflect exactly what the contractor has installed and where it is located. Have your draftsman update the original drawings and reissue them "As–Built."}

5. Correction of Work

a) Contractor shall, as Owner's representative directs, promptly repair, replace, or reimburse for any portion of the Work or any property of the Owner or of others that is lost, damaged, destroyed, or injured in any way by the Contractor while performing the Work.

Guideline 126

{When the contractor damages the work of others, such as of another contractor, you will have to get involved as the contractors do not have contracts with each other and there is no sure way of the injured contractor collecting. You are the common denominator. You can subtract money from progress payments to pay the injured party.}

6.   Telephones, Photocopying, Etc.

a)   The Owner will not provide telephones or photocopiers for Contractor's use. The Contractor will be billed for any telephone calls and photocopies made on the Owner's equipment.

b)   Contractors will provide their own radio communication system.

> **Guideline 127**
>
> {For vendors who are on site for short periods, you may want to supply the above as it is not feasible, nor cost effective, to expect them to supply the it. If the vendor is an offshore vendor, it is unlikely they will understand how to make the arrangements to acquire what they need.}

7.   Stores

a)   The Contractor will not be permitted to use the Owner's regular tool crib or safety stores.

> **Guideline 128**
>
> {For common items you do not want the contractor getting material from your stores. For hard to find or special items, you may want to supply them. An emergency situation would be another time when you would use the plant stores.}

8. Communication

a) Communications between Owner and Contractor will always be carried out between Owner's designated representative and the Contractor's superintendent or designated representative. Should Contractor incur costs at the instruction of anyone other than the Owner's designated representative, these costs will be charged to the Contractor's account.

Guideline 129

{The lines of communication have to be set at the beginning of the project and all contractors advised accordingly. In plants you will always have someone other than your designate trying to get the contractor to do something. Advise the contractors at your pre-award meetings that you will not accept any costs associated with this type of work request. You will have at least one or two instances of this happening, but do not give in and pay.}

9. Cleanliness

a) The Contractor's work areas must be kept clean and tidy and free of tripping and slipping hazards at all times. All cables (wire-rope, power, welding, etc.) and hoses must be coiled and stored if not in use or elevated and secured above floor level when they must cross traveled or work areas.

Guideline 130

{This tripping hazard section should be reinforced on a continuing basis, as it is too easy to leave welding cables and extension cords lying around. If you have plant personnel walking through the construction areas, the passageways should be clear.}

b) The Contractor must remove from site all trash and debris connected with their work. Only dry material can be dumped in the Owner's dry landfill site.

Guideline 131

{Check what the contractor's are allowed to dump in the plant's landfill site. You will get the complaints that one contractor was using the other's garbage bins or that "my guys never make a mess so this should not apply to me." It may be worthwhile to hire a contractor to do the entire cleanup and backcharge the contractors prorated on the number of people on site. This would not apply to concrete, etc.}

c) Contractor to keep the area around his trailer clean and on demobilizing to remove all trash and debris.

Guideline 132

{Most contractors are good about keeping their work areas clean. Keep watch for food waste that is not removed as it will attract rodents and wildlife. This will create a whole new series of problems for you if it gets out of control.}

c) Failure to clean up the site will result in the clean-up being done by others and the Contractor being backcharged accordingly.

Guideline 133

{Always make sure you still owe the contractor money at the end of the job. Then, if you have to pay someone to do the clean up it does not come out of your pocket.}

d)   Contractor shall also clean and maintain work areas in a tidy condition at the end of each working day so that existing Plant operations can proceed unobstructed during the night.

Guideline 134

{You do not want material spread around creating a fire or safety hazard. No matter how hard you try, plant employees will be in the work areas snooping around so it should be safe. This clean up also includes keeping the contractor's equipment out of the way after hours. There should be a parking area for the equipment and the keys be should not be left in the vehicles.}

e)   Construction areas are to be cleared of combustible trash daily.

Guideline 135

{To prevent fires the combustible trash should be removed. Even if removed, you should make sure that after hours, the plant security patrols through the construction area on a regular basis.}

10.   Use of Site by Contractor

a)   Only those persons authorized by the Owner's representative will be allowed access to the site.

Guideline 136

{You want to control who comes onto the plant site. All visitors should check in at reception, and there should be signage directing visitors to do so. If the contractor has employees you do not want on site, refuse them access. If you are not sure if someone has the right to be on the site, get advice from management. It's your plant and you control access.}

b) Contractor shall confine his equipment, the storage of material, and the operations of his workmen to the limits indicated by law, ordinances, permits, or the direction of the Owner's representative, and shall not unreasonably encumber the site.

Guideline 137

{Since the contractor's personnel will not know of the safety issues in your plant, it is imperative that they not be allowed to wander around the plant. If they have hard hats of a different color than the plant, you can easily spot them and direct them to their work area. If it becomes a big problem, have the troublemakers removed from the site.}

c) Contractor shall apply with Owner's representative for each vehicle required on the site for work of the contract. Owner's representative will provide approved vehicles with an identification card and all vehicles will be checked on and off at the gate by the Owner.

Guideline 138

{You want to control the number of vehicles on site and limit the plant's liability for accidents. The application should state that the vehicles are on site at the owners' risk. The plant is not responsible for any damage to vehicles. You also want a copy of the vehicle license numbers and a record of whom the vehicles belong to. You will always get someone parking a vehicle in a driveway blocking traffic, and you want to know who it belongs to so you can get it moved.}

Case History 15

We were building a greenfield plant and were at the cuts and fills stage of the site preparation. The area we had was quite large and we only had the earthworks contractor on site. We had a vehicle permit procedure in place, and as the contractor's large scrapers were parked in a remote area of the site, we issued permits to the three operators so they could drive their personal vehicles to their equipment. One day the local police paid a visit to the safety officer while investigating an insurance claim regarding a vehicle accident on the site. Apparently, one of the scraper operators wanted something out of his car so he drove over to the parking area, parked his scraper, and was in his car when he saw his scraper rolling towards the car. He got out of the way but as the three vehicles were parked beside each other, the scraper just pushed them together damaging them. The safety officer showed the police officer the vehicle permits stating that the vehicles were there at the owners' risk. The officer was satisfied that we were not libel and left us alone. In the end, the contractor was found libel and ended up paying for the repairs.

d)   Parking of Contractor's employee vehicles will be in the area designated by the Owner's representative.

Guideline 139

{If you do not designate a parking area, the contractor's employees will be parking in the plant lot. There will be complaints from the plant employees as the contractor will be working different hours than the plant creating parking problems, so it is better to have a separate parking area. If you live where you have to plug in your car during the winter, you should put in plugs in the contractor's parking lot.}

e)   The Contractor's employees shall enter and leave the site through the entrance designated for their use and shall be restricted to the areas on the site in which they are required to perform their work.

Guideline 140

{By using a separate entrance for the contractor you can control the effects of a contractor strike. The contractor would only be able to picket the construction entrance and not the plant entrance.)

f) Contractor will be responsible for transporting to and from the site of the work all Contractor's and Subcontractor's employees and all other persons the Contractor requires during the performance of the Work.

g) Owner does not provide job site security for Contractor's tools, equipment, and material. Contractor should supply his own lockable toolboxes and is responsible for all losses of tools, equipment, and material.

11. Protection of Property

a) Contractor shall take precautions to ensure that all cutting, digging, or other operations, which he is required to perform, shall not endanger existing structures, equipment, and facilities.

Guideline 141

{If your plant has an excavation permit or welding/cutting permit requirement, this should be stated and the forms attached.}

b) Contractor shall not cut or alter the work of others or the property of Owner without the written consent of Owner's representative.

c) Contractor shall provide suitable lockup facilities on the site for his tools and equipment and for those of his employees.

d) No tools, equipment, material, or other articles may be removed from the site unless the person removing such articles can clearly identify each article and establish his ownership and his right to remove them from the site.

e) To facilitate identification and the right to remove any article from the site, all articles shall be identified and checked in with the gatekeeper when taken onto the site, or, if taken onto the site by truck or rail car, invoices shall be retained and produced when requested. All tools and equipment shall be marked for identification and when taken onto the site, a form supplied by the Owner's representative (see attached) shall be filled out in duplicate and one copy shall be left with the gatekeeper.

List of Tools and Equipment

Item	Item

Date of Entry: date tradesman brings tools on site   AM/PM
Contractor: name of contractor
Contractor's Signature: signature of contractor
Guard's Signature: signature of guard on duty
Date of Exit: date tools are removed from the site   AM/PM
Contractor's Signature: signature of contractor
Guard's Signature: signature of guard on duty

A Removal Permit must be obtained for all items being removed before completion of Contract.

Guideline 142

{The above form is only used when the tools and equipment first arrive on the site and at project completion when the contractor is demobilizing the site. For those times in between, the plant should have a Removal Permit system in place, and this Removal Permit should be used.}

12. Safety

a) All work shall be done in a safe and professional manner in accordance with the Occupational Health and Safety Act (latest edition) and Owner's safety procedures.

b) The attached Contractor safety procedures shall be followed at all times.

Guideline 143

{The government safety acts are general in nature and your plant should have a site-specific safety policy or procedure. Include it in the tender document.}

c) Fall protection is required for all jobs performed at heights of more than 3.5 m (10 ft). There are **NO** exceptions.

d) Smoking is allowed only in designated areas.

Guideline 144

{Smoking is a problem both as a fire hazard and because of foreign material getting into your product. Set up the designated areas, such as vehicles and lunchrooms, and enforce the regulations with the contractors.}

e) Work permits must be obtained and filled out before any work takes place. Additional permits will also be required for excavation and hot work.

Guideline 145

{Some plants have a work permit system where the plant's operators have to issue permits every morning to the contractors. This lets the operators know who is working in their area and for how long. In case of emergency, plant operators know who to notify on site.}

13. Plastics Control

a) The Contractor shall comply with the Owner's Plastic Control guidelines and directives including those set forth in Schedule I, part A (attached).

> **Guideline 146**
>
> {If your plant has a cleanliness issue where it is important to keep out extraneous material, you should modify the following Plastics Control procedure to meet your needs. This procedure was developed for a plant where plastic of any type would affect the final product. In this case it was usually the customer who found the plastic.}

14. Setting Out the Work

a) The Contractor shall locate the parts of the Work on the site from the bench marks and basic reference lines established by the Owner's representative.

b) The location of each part of the Work shall be established by the method called for on the Construction Drawings and by no other method. For example, if the location of a part of the Work is established on a Construction Drawing from the centerline of a piece of equipment, the location of that part on site shall be established from that centerline and not from a bayline or the mid-point or face of a column or from some other physical feature.

> **Guideline 147**
>
> {Establish at the beginning of the work that you want the contractor to follow this rule. It makes it more difficult for the surveyor as he has to do more calculations, but at least the results will be accurate. If you look at large structural columns in a plant you will see that they are not machined pieces of material, but are bent and not square. So if you have

one surveyor running in a line off a column, the next surveyor may not use the exact same spot and will get a different answer. Use only actual centerlines.}

c) The Contractor shall not use scaled dimensions for the construction of the work. Should any dimensions be unobtainable from the Construction Drawings, the Contractor shall obtain them from the Owner's representative in writing.

15. Chart of Accounts

a) Contractor shall, before the first application for payment, submit to Owner's representative a chart of accounts (from the progress payment form) of the various parts of the project, including quantities and amounts aggregating the total sum of the Contract, made out in such a form as Owner's representative and Contractor may agree upon. Contractor's invoices shall be submitted in accordance with this chart and, if required, itemized in such form and supported by such evidence as Owner's representative may direct.

Guideline 148

{The usual process is for the contractor to make a preliminary issue to you for approval. If you agree to the progress and the amount of money then approve the issue for submittal. The contractor will try to get as much of your money in the fastest time possible. If you don't agree with the percent complete then change the amounts on the form and return to the contractor.}

16. Working Forces

a) The Contractor shall submit to Owner's representative and shall require each subcontractor to submit separately to Owner's representative before 9:00 a.m. of the next working day a Daily Manpower Report in a form satisfactory to the Owner's representative showing the number of people in each trade employed on the project for that day (see form attached).

Guideline 149

{You should get a daily manpower report from the contractor whether on his form or yours. For safety reasons you need to know how many people are on site each day, and you can use the reports to check FWO accuracy by checking to see if the proper people were on site to do the work.}

b) The Contractor shall employ tradesmen, labor and administrative personnel from the local area and district area to the extent that qualified persons are available.

Guideline 150

{Most plants have a local hire policy that applies to contractors as well. It is just good business for your contractors to hire local tradesmen.}

c) When Contractor employs personnel from outside the local area who are eligible for Living Out Allowance (LOA), the Contractor shall obtain proof of residence and be prepared to supply the Owner with such proof.

Guideline 151

{What you are trying to control is the tradesman who lives across the street from the plant and uses his brother's address, which is 100 miles form the plant, to qualify for LOA.}

17. Drawings and Other Documents Issued for Construction

a) The Work of the Project shall be performed in accordance with the latest revision to the construction drawings, specifications, and other information issued to Contractor at the site.

b) Drawings that Owner's representative furnishes to Contractor for the actual performance of the Work of the Project will be marked "Certified for Construction."

c) Three copies of each construction drawing will be issued to the Contractor at the site and up to three additional copies upon request of Contractor.

d) Construction drawings may be revised from time to time during the course of the Project.

e) If construction drawings, specifications, or other documents issued for performance of the Project differ materially from Contract drawings, specifications, and other documents as executed and such difference would result in a change in the cost in carrying out the work of the Project, then Contract Amount shall be adjusted subject and pursuant to the General Conditions of Contract Part 6 of CCDC No.2.

f) Some of the construction drawings may be issued with certain areas such as equipment pads or anchor bolts marked "HOLD" where a portion of the design is incomplete for some part of the Work of the Project which the Contractor is not required to do at that time. In this case the Contractor shall proceed with parts of the Work not marked "HOLD."

g) Two sets of all specifications and other instructions relating to performance of the Work of the Project will be issued to the Contractor at the site and any necessary additional copies of individual documents will be issued on request.

Guideline 152

{All drawings and specifications issued to the contractor are issued through your representative on site. You should not send drawings directly from the design office to the contractor. All drawings and specifications should be issued using a transmittal.}

18. Manner of Construction

a) Owner's representative shall have the right to review with the Contractor the manner in which Contractor proposes to perform or is performing the several parts of work of the Project. If, in the opinion of the Owner's representative, the proposed manner of performing the work does not insure the safety, accuracy, or proper rate of progress, Owner's representative may require Contractor, at no additional cost to Owner, to change the manner of performance of the work to one of which he does not object.

b) No requirements to change the manner of performance or lack of such requirements by Owner's representative shall relieve Contractor of his responsibilities for the performance of work of the Project in accordance with the Contract.

19. Acceleration of Work

a) If Contractor fails to complete any part of the work of the Project in the time specified in the construction schedule, or if it becomes apparent to Owner's representative that the Work will not be completed in the time specified in the contract, and if failure is due in whole or in part to any fault of the Contractor, Owner's representative may require the Contractor to expedite work on the Project. This may include, working overtime, using additional shifts, employing more men or more equipment, or by any combination of these required to accomplish the Work according to the construction schedule at no additional cost to Owner.

b) Owner's representative shall have the right to require Contractor to complete work of the Project or any part thereof before the date set forth in the construction schedule or to require Contractor to complete work of the project according to the construction schedule when his claim for delays has been ruled valid by Owner's representative. Contractor may submit a claim for reimbursement of any additional costs to him that can be shown to be the direct result of the exercise of such right by Owner's representative.

Plastic Control Guidelines

Schedule I, Part A

Guideline 153

{This is an example of the type of guidelines you can develop for your plant. In this case, plastic created a big problem if it got into the pulp process since it melted and fine particles did not come out in the process. If your plant has metal, plastic, paper, glue, or some other contaminant that has to be kept out of the final product, use the following to develop a guideline.}

Case History 16

In one pulp mill we were getting plastic contamination in the final product. There were already restrictions on plastic use in the plant. The cafeteria went to metal utensils. They changed the hearing protection to ear muffs and outlawed plastic combs. This cleaned up most of the problem; however, there was still plastic getting into the process but it was not consistent. As the plant used wood chips to produce the pulp, each load was sampled to determine payment. Someone noticed that one of the people doing the testing was eating his lunch on the chip piles and discarding his plastic sandwich wrappers into the pile. This was the source of plastic that was giving the problem. This points out that you have to look at every detail to keep contaminants out of the system.

Plastic type material can contaminate the pulp produced on this site; therefore, any synthetic material such as polyethylene, polypropylene, and Styrofoam must be managed very carefully by the Contractor to ensure that the material does not get into the pulp process.

As a Contractor working on site, you have the obligation and responsibility to minimize the quantity of harmful plastics entering the site and whenever these contaminants are used, careful usage and control must be exercised and proper disposal in suitable containers must occur.

The following are SOME guidelines involved in successful plastic control:

1. Roping off areas - Polypropylene rope (yellow in color) will not be used. Jute or hemp rope will be used.

2. Slings - Polypropylene slings are banned. Use steel cable, hemp, or nylon slings as alternatives.

3. Signs - If plastic signs are used, they must be securely fastened to ropes, walls, sign posts, etc. to prevent the wind from carrying them away and into the process.

4. Consumable items - These must be disposed of as soon as possible. Examples include oil and coolant containers for portable welders and compressors, welding rod packing, etc.

5. Packaging - The mill stores will remove as much plastic packing as practical.

6. Employee items - Combs, ball point pens, thermos bottles, lunch kits, and associated items are serious problems if they enter the system. Employees must be aware that saran wrap, baggies, etc. should be carefully discarded in garbage cans. Employees can also help by using pencils instead of ball point pens and by using "nylon" combs.

7. Styrofoam - Cups, insulation, packing materials, etc. should not be used.

8. Safety items - Hardhats, hearing, nasal, and eye protection devices can be potential plastic contaminants. Our Safety Department can assist you in obtaining acceptable safety items which meet our standards.

9. There are numerous other sources; therefore, vigilance is the key.

This is the end of the Tender Document section and the document is now ready to be issued. The document is issued by the purchasing group as they will add the plant terms and conditions. The next step is the bid meeting, which is referenced in the tender document, the bid evaluation, writing of the purchase order and the contract. These issues are covered in the next chapter.

Chapter

8

BID MEETINGS, BID EVALUATION, PURCHASE ORDERS, AND CONTRACTS

Bid Meetings

For all contractors to have the same understanding of the extent of the work you should hold a bid meeting on site that includes a visit to the location of the work. On site bid meetings are normally held with equipment vendors. There is a section in the Tender Document for advising of the time and place of the meeting. You cannot force a contractor to attend but it is usually advised that they do. Unless you are pressed for time, allow at least two weeks between sending out the bid package and the bid meeting. This will give the contractors reasonable time to make arrangements to attend your meeting. If you are pressed for time and your contractors are local, you can have the meeting within a week after issuing the bid package. A few days before the meeting send a fax requesting a response from those who will be attending the meeting to get a feel for how many will be there.

Make sure you have a meeting room reserved and if there are any meeting location changes advise the contractors in writing before the time of the meeting. Advise your plant security of the location and whom you expect will attend. Give him the location of the meeting so he can give proper directions to the contractors; otherwise, they may end up wandering around your plant. If the contractors need their vehicles or special safety equipment for the site visit, advise the security guard.

Arrange for an assistant to take minutes at the meeting so you can concentrate on the discussion at hand. You also may need a second opinion in case something comes up later about the meeting. Have a sign-in sheet and ask everyone there to sign their name, print it out, and print the name of the company they are representing. Keep this list for future reference. Start the meeting on time, as it is not fair to those who are punctual to have to wait for those who are late.

Have an agenda for the meeting outlining what will be covered in the formal meeting and in the site visit. At the meeting start at the beginning of the Tender Document and go through every section. Review what you have asked for and what you expect to get back from the contractor. Ask if they understand what you want and if there are any questions. You do not have to have all the answers at the tip of your fingers. You can advise the contractors that you will get back to them, in writing, within the next few days.

Review the drawings, standards, and specifications at the meeting. If you have additional or revised drawings, issue them to the contractors at the meeting and review what the changes are. After the drawing review, go through the Special Conditions and make sure the contractors understand what services and utilities you will and will not provide. Any misunderstanding in this section can result in an unwanted surprise to the contractor, which you do not want.

Once you have covered all the written issues, drawings, and specifications and there are no more questions, it is time to take a site tour to see the work location. Take note of who is on the tour. On the tour make sure you point out the locations of the services such as water, air, power, washrooms, trailer area, and parking. If it is a greenfield site, explain what the set-up will be and where the above noted services will be located on the site. Take note of any questions answered and include them in the meeting minutes. Make sure all contractors understand the site conditions and any hazards the site could impose on them.

When the meeting has finished it is imperative that the minutes be written and issued within 24 hours. These minutes will form part of the contract documents. You can include clarifications in the minutes or send them out under separate cover as an addendum. If questions come up between the meeting and the bid due date, respond in writing to the pertinent people. These responses will form part of the contract. You will get requests to extend the bid due date. If all contractors are requesting an extension, you may not have allowed enough time for the bid to beprepared. You have two choices:

- If you are pressed for time with your schedule, you can refuse to extend the due date;
- You can extend the due date as you see fit.

If only one contractor wants an extension, you will probably have to deny the request.

The contractors' quotes and equipment vendors' quotes should be sent to the purchasing group where they will be collected. They are date stamped when received and collected until the time of the opening. If you receive any, you should give them to the purchasing group unopened. Make arrangements with the purchasing group to meet when all the quotes will be opened. Once opened, they have to be compared to determine the successful bidder. This comparison is called the bid evaluation.

Bid Evaluation

The bid evaluation process is used for both contractors and equipment vendors and involves comparing all the quotes in a systematic and organized fashion. The bid evaluation is the lead-up to a pre-award meeting which:

- will be your last chance to make changes, request exactly what you require, and have this included in the price.
- is the opportunity to ask questions and clarify your understanding of what is included,
- allows you to negotiate quoted prices you may not agree with.

The evaluation should be done in a spreadsheet format where items can be compared across all bidders. It is important to do this comparison on a spreadsheet as it:

- makes the differences between bidders easy to see;
- allows those who require specific information to get what they require very easily.

Because most of the information in the bid evaluation will appear on the purchase order, be as organized as possible so you only have to sort out the information once and then copy it to the purchase order. The bid evaluations should be considered confidential information and treated as such. Do not make them a general distribution document.

The following items should be included in the bid evaluation. (Their order is not important.) The items apply to both equipment vendors and contractors. This exercise may seem trivial for some small bids; however, no matter the size of the bid, the same amount of work has to be done. Only by doing this for the small projects do you get the experience required for the bigger projects.

Price Inquiry Number

This allows you to reference the correct bid package.

Price Inquiry Name

This lets others not familiar with the numbering system know what bid you are evaluating.

Date Evaluation Done

For reference purposes only. If you do not record this, you will forget when you did the evaluation.

Contractor/Vendor Information

Show complete information such as name, address, phone number, fax number, contact person. This lets you and others get quick access to the vendors. The contractor/vendor quote number and all relevant faxes, memos, letters, and telephone conversations that apply should be noted. These are identified by date.

Technical Evaluation

Cover all technical items that are applicable and highlight any differences. This can take several pages depending on what is being evaluated.

- Are codes and standards followed?
- Have they followed the Standard Component List?
- If labor rates are given, list typical ones and indicate how other contractors compare in relation to this one.

Pricing

Items to consider:

- Are funds in your country's currency?
- Is exchange rate applicable and if so what rate?
- Is price firm or subject to escalation?
- How long is the quote good for?
- Are there unit prices for additions and deletions?
- Are taxes broken out?
- FOB point?
- Freight allowed?
- Duty and brokerage included?

Terms

Look at the terms of payment:

- For equipment, are there progress payments?

- Are equipment lump sum payments based on milestone dates?
- Are there exceptions to any of the plant's terms and conditions?
- Are equipment payments subject to holdback?
- Are the contractors looking for a lump sum down payment before they start work?
- Are contractors looking for a mobilization payment? How much money is held back until contractor has demobilized?

Delivery

Check the turnaround on drawings. The contractor/vendor should allow 10 days for drawing review in your office. If you are aware of any present or potential delivery restrictions, make a note to question the supplier. If material is being shipped out of Europe, certain countries shut down for the summer months and this could affect your drawing reviews and delivery.

- What is the expected equipment shipping date?
- Is this after receipt of approval drawings?
- How much time has been allowed for drawing review?
- Does it meet the plant requirements?
- Where is the delivery being made from? North America? Europe?

Shipping

If you are aware of any present or potential shipping problems, they should be noted and questions asked of the supplier. In the colder parts of the Canada and the United States, there are load restrictions on the highways. You may have to find alternate routes.

- How will it be shipped?
- Shipping weight given?
- Shipping dimensions given?
- Shipping restrictions such as temperature, size, weight, etc.?
- Suitable for outside storage?
- Packaged for ocean shipping?

Plant Project Engineering Guidebook

Warranty

- What is the mechanical warranty?
- When does the warranty period start?
- What is the performance guarantee?
- What has to take place to access performance?

Union Affiliation

- What unions will be involved in the contract work?
- What is the union contract expiration date?
- What happens if contract expires during the contract period?

Startup Services

If you are using a European supplier, you may have to provide them with an office, phone, and fax. Make a note to add this cost to the final total.

- What is provided?
- How many days included in their quote?
- Does it include travel and expenses at cost?
- What is rate for additional days?
- Is training included or extra?
- Who does it?
- How is it done?
- What costs do they expect you to pick up?

Certificates

European vendors do not have workman's compensation programs. Most companies have straight insurance policies to cover their workers when on international travel. This insurance is not recognized by worker's compensation. Some vendors who work in North America have a worker's compensation account but not in every state or province. Make a note to question any European vendor about the worker's compensation coverage. Check with your plant safety group, as you may be able to cover them under your plant's policy.

- Are certificates for insurance, workman's compensation, etc. available?

Spares

- Are recommended spares listed?
- Are the spares included in pricing?
- Is additional pricing given for spare parts?

Vendor References

- Are references given with contact names?

Local Content

- Are local content requirements, if any, being met?
- If not, state so.

Area Representation

- Who is the closest representative?
- Where is the closest warehouse?

Below is a checklist developed from the above. This checklist can be used on an individual as-needed basis and is not necessarily a record-keeping document. Its purpose is to help identify those items that should be incorporated in a bid evaluation, although not all items will apply.

This checklist can be used for evaluating bids from equipment vendors and contractors. In going through bids, you may find that some items are in one bid but not another. By using one checklist for each bid you can quickly check to see what additional information is required from the vendor or contractor. Those items that do not apply to the particular bid should have NA in the checkoff box. Using this checklist will ensure that your bid comparisons are complete and are all based on the same information.

When comparing contractors' bids for equipment/piping installation, you may find that their Additions/Deletions and labor charge-out rates will vary with one contractor being high in some areas and low in others. To do a comparison, make up a sample "Contract Extra" including at least one item from each Addition/Deletion. Using this "Contract Extra," cost it out using each contractor's "Additions Cost." You may find that they are all fairly close in the total cost or one may be too high in comparison with the others. You should then look at the project work and make a judgment call as to where you expect the bulk of your extras to come from and reassess the costs. If the contractor with the high extras is otherwise competitive, you may want to go back and tell him to rework his extra cost to make him competitive.

Look at the value you are getting for Deletes. They are always less than the Additions. If you do not agree with them, talk to the contractor about getting more money back for Deletions. You may not be able to increase the values across the board, but the contractor should move on some of the Deletions. All negotiations for the Additions and Deletions should be completed before contract award. This negotiation is usually not done, and once the adds and deletes start applying it is too late to negotiate. Do it while you have the chance or you may regret it later.

Bid Evaluation Checklist

1. Price Inquiry No. Shown	**8. Delivery**
2. Price Inquiry Name	Date of Shipment
3. Date Evaluation Done	After Approval Drawings
4. Vendor Information	What Will Be Supplied
complete address	How Many
name of contact person	What Type
phone number	What Size
fax number	Match Vendor Date Req'd
quote number	**9. Shipping**
5. Technical Evaluation	How Shipped
Mechanical	Shipping Weight
Piping	Shipping Dimensions
Civil	Humidity/Temp. Restrictions
HVAC	Suitable for Outside Storage
Electrical	**10. Warranty**
Instrumentation	Warranty or Performance
Codes and Standards	**11. Startup Services**
Highlight Differences in Bids	Number of Days Included
6. Price	Includes Travel and
US or Canadian Funds	Expenses at Cost
Firm or Subject to Escalation	Rate for Additional Days
Exchange Rate	Training Provided
Unit Prices for Add & Deletes	How
Taxes Included	Freight Allowed
(break out if included)	**12. Union Affiliation**
FOB Point	Contract Expire Date
Duty and Brokerage Included	**13. Certificates Provided**
7. Terms	**14. Spares Included**
Progress Payments	Price of Spares
Milestone Payments	**15 Vendor References**
Subject to Holdback	**16. Local Content**
	17. Area Representation
	Closest Warehouse

Once the evaluation is complete you should find that the numbers are fairly close. One price will be high, one will be low, and the rest in the middle. The high priced bid could be that way because:

- they did not understand the work,
- they are too busy and do not want the work but responded to stay on the bid list,
- they are just a high cost company.

The low priced bid could be that way because:

- they did not understand the work and have left something out,
- they have intentionally left something out and are trying to get to the table where they will bring in the rest of the costs,
- they are a low cost company.

You do not have to take the lowest price. If you can honestly justify not taking the lowest price, then don't. Do not let anyone rebid after the bids are open. If you let one rebid, you will have to advise all the others that they should rebid as well. You will always be asked, but do not tell other vendors/contractors what the bids were or how much higher they were above the lowest price. This is confidential information.

Purchase Orders and Contracts

Before any purchase order or contract is issued, a pre-award meeting should be arranged with the selected vendor or contractor. For a straight equipment purchase this is not necessary; however, for a supply and erect or erect only contract, it is highly desirable. This meeting is to review the complete contract package and to make sure you and the contractor agree on what is in the contract package. You should have a meeting even if there have not been any changes to the bid documents during the evaluation phase.

The pre-award meeting is your last chance to make changes and get those changes included in the lump sum price. After this meeting everything is an extra. If dealing with a cost plus contract, you should review the scope of work to make sure there is a clear understanding of what is required.

If you have questions for the contractors or vendors, remember that it may take some time to answer them. Therefore, it is advisable to ask them, in writing, the time-consuming questions before the meeting is held. That way the contractor or vendor can come to the meeting prepared to answer your questions.

The meeting should be a complete review of all the documents to make sure all the faxes, letters, phone conversations, etc. are included; that all bid drawings are agreed to; and all standards are agreed to. (The minutes of the pre-award meeting will become part of the contract package.) Sometimes negotiations will take place to try to get a better price. When this happens keep in mind that "There Are No Free Lunches." If the bid value is reduced, the money is coming from somewhere and it is usually your pocket. The money will be made up through reduction in scope, changes in material specification, and/or contract extras. You will find out eventually, but by then it will be too late to do anything about it.

Once a purchase order is awarded, a minimum of two sets of all the contract documents should be signed and dated by you and the contractor. One set is yours and the other set is the contractor's. The documents include all the drawings, scope of work, standards, specifications, faxes, minutes of meeting, etc. These signed documents form the basis of the contract and should be kept in a secure location so they do not get damaged or destroyed. You will have to refer to them often.

You should come to some agreement with the contractor on the progress payment breakdown. If you do not agree and feel that the contractor is front-end loading, get the values changed at this pre-award meeting. (A contractor will look at the total work and pick which work will be completed first. Those items will then carry the highest value on the progress payment form, which results in the contractor getting the majority of his money at the beginning of the project. This is what is meant by front end loading).

If, as a result of this meeting, the value of the contract changes in total value, have the contractor change the progress payment breakdown. Do not do this breakdown for the contractor, even if he asks you to.

When contractors put a bid together they may take your RFQ apart. To make sure the contractor has all the relevant forms, reissue them to the contractor using a transmittal or letter. You can make up a disk of the progress payment forms, filled in correctly, and give this to the contractor as well.

Each plant has its own policy about who awards purchase orders and contracts, but the purchasing group should be involved. The plant should have standard contract documents and, if need be, the purchasing group should get the contract prepared and assembled by the legal department.

You or the purchasing group should send a letter to all the unsuccessful bidders advising them of the fact. You do not say who got the contract (although they probably know anyway). This is the professional thing to do.

Any purchase orders or contracts should be written up as soon as practical—the sooner the better. Remember that the accounting group needs a written purchase order to pay invoices against. When writing the purchase order, it should include the following details:

- Complete address of vendor with contact name, phone and fax numbers.
- The reference documents with numbers and dates so they can be accurately identified.

- The total cost of the purchase order with a cost breakdown assigning costs to the appropriate cost code for accounting purposes. The total cost should be written out as well as numerically expressed.
- Detailed descriptions of exactly what you are purchasing. This should include a technical description of the items, delivery dates, painting instructions, erection and startup supervision requirements, process and performance guarantees, tagging information, and shipping instructions, if necessary.
- Other information relevant to the purchase order.

The above work may seem unnecessary when all you have to do is attach all the documentation to a short, written purchase order stating "See Attached Documents." The problem with not writing everything out is the "Attached Document" does not get to the people who require it and no matter what you do, it is a very difficult problem to correct. Attached sheets get lost and sometimes never make it through the system. Only by writing out all the information can you ensure that:

- The vendor knows exactly what you are buying;
- Anybody replacing you knows exactly what you bought;
- The stores department knows what is being shipped and whether or not it is complete when the order arrives;
- The accounting group knows what they are supposed to be paying for;
- Your field crews know what they are looking for, what is being provided, and who the contact person is in case of problems;
- For record keeping purposes (files), all information is in one document.

Your intent is to make sure everyone involved has the same information. Size of the order does not matter as mistakes can be made with large and small orders. You should be able to pick up the document at a later date and know exactly what you ordered. This can only be achieved by writing out all the information.

Once an order has been issued there should be a mechanism in place to handle changes to the purchase order. With vendors it is usually a revision to the purchase order and with contractors it is usually fieldwork orders (FWO) or a contract change order (CCO). (The latter two will be discussed under Construction Management.) The plant should have a procedure in place to handle both of these situations.

When revising a purchase order for a vendor, it is done the same as writing out the original order. The additional information required on a revision is the same format as the original purchase order. You obtain the new total value of the purchase order by showing the following on the document:

Original PO Value:	$xxxxx.xx
Additional, This Revision:	$yyyyy.yy
New PO Total:	$zzzzz.zz

If another revision is necessary, the Original PO value would be the previous New PO Total ($zzzz.zz). It would now be called the Previous Revised Total. By adding up the new total for every revision, the PO value is readily available to the interested parties. You are probably the only person in the organization who knows how the final dollar value was reached, so it is up to you to keep everyone concerned informed.

This is the end of the section. However, this is not the end of the procurement function in the project. You have awarded the contract and it is now time to move out into the field to manage the work. You can now put on your Construction Manager hat as we get involved in construction management.

Chapter

9

CONSTRUCTION MANAGEMENT

Where Does it Start?

When the concept of construction management is mentioned, most people think of it as directing contractors on and around a plant or construction site. They believe that once the purchasing group has awarded the contract, a construction manager is appointed to oversee the administration of the contract. In many cases this is true; however, construction management generally involves more than this.

Construction management starts in the project design phase and runs throughout the project to its final acceptance. As the construction manager you should be involved in the design, procurement, and construction phases, i.e., throughout the project life cycle.

If it has not already happened to you, there will come a time when you will have to work on a project where someone else has done all the up-front work and you have been left to pick up the pieces and make a success of the project. Not being involved in all phases of the project is one reason for project failures. So, for a reasonable chance at project success, you should be involved in all of the up-front work. You should be looking at the following:

- Does the layout allow the equipment to be installed at the least expense? Working in existing plants limits the available room for new equipment installation, so the location of the equipment can have an affect on the cost of the project. If you are working in the middle of a building, a helicopter may be needed to install equipment; however, relocating the equipment to an outer part of the building may allow a crane to be used. Sometimes moving a piece of equipment 10 feet will change the type of equipment required for installation.

- Is the design constructable? To play it safe, assume that the designer has limited knowledge about construction or what happens on a construction site. You may get lucky and get a designer who understands the field issues; otherwise, the designs produced will look good on paper but will not take into account others working in the area, existing items that may be in the way, or the sequence of installation. Sometimes if the designer does not know what to do, he will just leave the item out. (When in doubt, leave it out?) It then becomes a field problem. If the designer is lucky, he will be on another project before the field realize there is a problem. On large projects there can be many bombshells from this lack of field knowledge. The problems will come at you from all directions and at the least opportune time. Make yourself available for design reviews, as it will save you a lot of trouble in the long run.

- Does the design take construction safety into account? You will get a lot of designs where there has been no thought given as to how to install the equipment. Review what is being proposed and see if the installation can proceed safely. Using a helicopter for moving material, etc. sounds like fun, but in an operating plant it can become a problem to organize it so that the procedure is carried out safely.

- Is there room to maneuver around the equipment? Sometimes layouts are made so tight that installation and maintenance can be difficult. Always be prepared to maintain the equipment during construction, as there may be a problem with it. Equipment problems during construction are usually more involved than simple maintenance. Review the layouts and make sure there is enough room to maneuver around the equipment and to get components out.

- Is there a method in place to minimize interference with equipment and services? If you ask a pipe designer, an electrical designer, and an H&V designer to run their respective material through a wall, they will, acting independently, usually pick the exact same spot. Even with CADD there is no guarantee there will not be interference. Check and see what system the designers use to minimize interferences. This interference issue should be part of a design review.

- What type of contracts will be used? The type of contract you plan on using can affect the design phase. If you are using a lump sum contract, the drawings have to be fairly detailed. If using cost plus there is less detail in the drawings but more definition in the contract documents. It helps to define these issues at the beginning.

- How well written are the scopes? You should be involved in the writing of the scopes. The designers will assume the work splits between contracts and the work splits will differ from what you really want. Based on your understanding of the field, you will want to arrange the work scopes over the different contracts.

- Are the methods of payment for excavations and backfill clearly defined? There are different methods of calculating these unit prices, so make sure you have one that works in your situation.

- Additions and Deletions - Based on your knowledge of the work and the scope, are there adequate additions and deletions listed in the appropriate section of the tender document? Try to cover as much additional work as possible. You may not get the best price for additions once the contract is signed, work is underway, and the contractor realizes that he is loosing money on the work.

- Who is on the contractor bid list? Include only those people you want on the bid list. If in doubt about anyone, leave them off. Unless you know the contractors, you should investigate them all.

- Progress Payment Forms - Are the progress payment forms made up in a reasonable manner that allows tracking with the project budget and for any other tracking required? Since you are managing the budget, the progress payments should be based on an arrangement that suits your budget format.

- What transpired at the bid meetings? Are minutes available? Who was there? What really went on? You should attend all bid meetings. Things will transpire at these meetings that will not necessarily be in the minutes.

- Get involved in the bid evaluation. The bid evaluation process used should result in the most economical contractor being selected. You should have some input from an installation and contract administration point of view.

- How is the contract awarded? Who writes the purchase order? Confirmation may be by fax. Make sure whoever writes the PO sends a handwritten copy, complete with backup, to the field.

- How will the site be organized? This is your responsibility, and several months before actual construction you should start looking for site people. Having some of the designers on site as field engineers is a big help.

Construction Management

The size of the project and the experience of the plant personnel will determine how the construction management will be handled. For large projects, where the plant personnel have little experience at managing, a consultant can be hired to do all the management (project and construction). This is an extension of the design process and can be a separate contract; i.e., have one contract for the design and another contract for the construction management. The construction management can go to another consultant if he proves to be more experienced. This is quite common today. With this set-up the design consultant would design the plant and write/issue purchase orders on his own letterhead and the construction management consultant would organize and manage the site. The plant would have a person to keep track of the project and to work as liaison between the design consultant, the construction management consultant, and the plant to make sure the project runs smoothly.

If the project is small and the plant personnel have the experience, the plant can handle the project and construction management. Get a consultant to do the design and have him write and issue all purchase orders on plant letterhead. For plant personnel construction management can be a very time-consuming exercise and this creates a potential for problems, as there never seems to be enough time for adequate contractor supervision.

General Contractor

The above distinctions are important in determining who is the general contractor—the plant or a consultant. If a consultant hires the contractors on his purchase orders and is supervising the site, then he is the general contractor. If the plant issues the purchase order to the contractor, then the plant is the general contractor even though an outside consultant is hired to supervise the contractor. You have to be aware of who the general contractor is, as the general contractor is liable for the project. This means he is responsible for site safety, fist aid, security, and directing all contractors on site who are associated with the project. Generally, on small projects plants do not have a problem with the risk and are prepared to assume it. On larger projects it is too much risk for the plant to assume.

The biggest problem with being the general contractor is site safety. This includes responsibility for ensuring that there is a safety program in place and that the appropriate safety measures are being followed. This responsibility cannot be delegated to the site contractors doing the work. The contractor has to have his own safety program but it is specific to his trade. The plant must have a safety program specific to the plant. To protect the plant in case of a serious accident, you will have to show that you, as the general contractor, performed due diligence by being proactive with the safety measures. To protect yourself, notify the contractor in writing of any infractions, and point out to the contractor's personnel any time they are performing an unsafe act, etc. All conversations and actions taken regarding safety issues should be in writing and filed for future reference. Your best protection is to hire a safety consultant, trained in construction safety, to handle this issue for you.

This issue of safety is a big responsibility especially on a plant construction site. What is safe or unsafe depends on your perspective. If you do a tour with a construction safety professional, you will see that he looks at work practices differently than you do. He will pick out the infractions that you will miss because you are not trained to look for them or you are not familiar with the rules. You have to be vigilant and if you are not sure, question the act being performed. Do not turn your back on safety infractions and hope for the best. If there is ever a problem, you want to be able to prove to the authorities you were proactive and did what you could to make the contractor aware of safety issues, infractions, and errors in implementing his and your safety program. Always have the health and safety of those under your direction in mind when they are performing their job and discussing ways of doing jobs. You do not want a serious accident on your job site with the authorities holding you responsible because you were not diligent.

Contract Signing

Once all the bids are in and have been evaluated, it is time to make a decision. The short list contractors are selected and individual meetings arranged to discuss the final price. At this meeting you should review everything in the scope to satisfy yourself that the contractor has included all items called for in the tender document and that you both have the same understanding of all the items. This is your last chance to have items and prices changed without the penalty of an extra. Get a copy of the contractor's Health & Safety Policy and the Quality Control Procedures before contract award. If you are meeting with several contractors, at the end of each meeting tell the each contractor you will get back to him and go on to the next contractor meeting. If the contractor is the only one on the short list, you can, based on your company policy, award the contract at the meeting. At any of these contractor pre-award and award meetings, a purchasing representative should be present. Once the decision to award is made the following has to be done, but not necessarily in the following order:

- Give the contractor written confirmation of the purchase order number. Copy the construction manager.
- Advise, in writing, the unsuccessful contractors.
- Determine who the contact people are for communication on the project.
- With the successful contractor, sign off the drawings, standards, and specifications that make up the quoted price.
- Make sure the progress payment forms are correct and issue them to the contractor and to your accounting group.
- Issue all relevant forms to the contractor.
- Advise others in the plant what has transpired and when you expect the contractor on site.
- Write the purchase order/contract.

Written confirmation of the purchase order number can be by a fax sent to the appropriate contact person or by written letter. It is important that the contractor have this information in writing, as some contractors will not make a move unless they have received written confirmation.

Sending letters to the unsuccessful contractors is the professional thing to do. If not done, the contractors will be phoning you for the results. The purchasing group usually sends out a standard form letter and it is a recommended practice.

The final set of bid documents have to be signed off to establish the base from which extras will evolve. Sepias of the drawings (if available) or two or more sets of prints are initialed by both parties and dated. The contractor gets a set for his files and you get a set for your files. The site construction manager, if you have one, will require one set. All signed off documents should be kept in a separate file and not mixed in with all the other drawings, etc. These are important documents and you will have to refer to them throughout the project.

When the bid documents come in, the contractor will have given a breakdown of his price based on the breakdown you requested. As the quotes are finalized and prices are changed, this breakdown may not get updated. After award get the contractor to redo his breakdown to reflect any price changes he has made or that resulted from discussions held. Advise him that he will not get paid until the changes have been made, as accounting will only pay based on this progress payment document. Once you get the breakdown, prepare an official progress payment form and issue it to the contractor plus others in your group who should get it. This is proprietary information and should not be widely distributed nor left lying around.

All forms mentioned in the tender document should be issued to the contractor. Make up a transmittal so you know the forms have been sent and received. This way there are no excuses for not using the forms.

Once a contract is awarded, inform the required people in the plant. Advise them as to who will be on site, when they will start, and where they will be located and working. The rest of the plant should be aware that there will be strangers in the plant and extra precautions may have to be taken. You do not want the contractor's employees wandering around the plant and getting into areas they should not be in. Have plant employees report any problems to you and then handle them appropriately.

The purchase order/contract should be written and put into the system as soon as possible. If you have site staff, this is the only document that describes exactly what the scope is. Keep in mind that any documentation attached to the purchase order will usually get lost in the paperwork shuffle. Therefore, you should send all backup documentation directly to your site personnel. Also, as it can take a while for the official document to go through the system you should issue a handwritten copy of the purchase order or contract to your site personnel.

If the information is not sent to the site, your site personnel will be working blind. In order to have control over your construction site you have to be the one with the information. Your site personnel should not be going to the contractor for information you are responsible for generating.

Contract Administration

The contract administration process actually starts in the design phase with drawings, specifications, and codes. The drawings should be as thorough and detailed as possible within in the time constraints available. The more information in the drawings, the better defined the scopes will be. There is a cutoff point after which more time spent on the drawings will not improve them substantially and you will have to determine when this point is reached. The specifications and codes should be reviewed to make sure they apply to the project, and they should be spelled out and referenced in the Tender Documents.

All communications with the contractor should come from one person in the plant and should be directed to the designated contractor's personnel. Communications with any of the contractor's subcontractors should go through the contractor as well. At the first meeting with the contractor, the contact personnel for both sides should be agreed upon. The contractor should be told that any extra work has to be approved by the plant representative, otherwise it will not be paid for. If this is not stated up front you will get billed for extra work requested by some person who happened to be in the area but the contractor does not know his name.

Field Work Orders and Change Orders

For additional work required over and above what is spelled out in the contract documents, use an Extra Work form or a Field Work Order (FWO) form. FWOs are extensions of the contract and become part of the main contract they are referenced to. Figure 9.1 is a standard Field Work Order form and can be modified to suit your jurisdiction.

Field Work Order

This work order is your authority to proceed with the work described herein. Should there be any question as to whether or not the work is within the contract or extra to the contract, a final decision will be made later by an authorized representative of the owner. All verbal requests to perform extra work must be immediately covered in writing on this form. For time and material work, labor and equipment time sheets are to be submitted daily. At the completion of the work covered by this work order, the contractor's invoice together with any necessary supporting documents are to be submitted to MHS Engineering Services Inc. for payment.

Guideline 154

{If you check your tender and contract documents, you will note that extra work can be priced as you determine and not what the contractor wants. That means extra work can be:

time and material based on contract terms;
lump sum pricing;
unit prices based on contract terms;
no cost}

Plant Project Engineering Guidebook

To: _____   Date: _____

_____   Field Work Order No.: _____

_____   Code of Account: _____

Cost/Estimated Cost: _____   Contract No.: _____

Contractor Reference No.: _____

Pricing Basis: Contract Work _____   Lump Sum _____

Unit Price _____   Time & Material _____

Other: _____

All Conditions contained in your contract with MHS Engineering Services Inc., referred to above, remain in full force.

Complete Description of the Work:
(Refer to drawing numbers where applicable).

_____
_____
_____
_____

Reason for Change:

_____
_____
_____

Agreed Start Date: _____
Agreed Completion Date: _____

Approved By: Approved By:

_____   _____
    accepted by addressee   project manager

Note: If this work is to be backcharged to a third party, insert the name of the third party here:

_____

Backcharge accepted by: _____

Backcharge No. _____

Figure 9.1 Field Work Order

With lump sum pricing the contractor will assess the risk (unknowns) he sees in doing the job and will include money to cover this risk, as he does not want to loose money. You will have to look at the problem and determine what risk is involved in the extra work, what the unknowns could be, and the problems trying to solve them. As an example, if you require the contractor to dig a hole, there may or may not be water. The contractor would try to include for pumping when it may not be required. You can opt for lump sum and if the contractor hits water you will then pay him for pumping on a time and material basis. You may want to discuss with the contractor what his concerns are about doing the job, but in the end the method of pricing extra work is entirely your decision. If there is no statement in your tender documents about extra work, have it added. It is another part of project control.

Since each FWO is a separate, identifiable document, keep a log of FWO numbers issued. The log should contain as a minimum:

- Sequential numbers
- Date issued
- Description
- Dollar value (estimate or otherwise)
- Has the FWO been written?

FWOs are not open ended with unlimited value. Your plant policy will have a defined limit which you are not to exceed without further approvals. For day-to-day work, most policies have a two level approval process. The construction manager is the first level and you are the second level. If the FWO limit is exceeded, you have to get management approval. The limits can vary, from $5,000.00 up to $65,000.00. It depends on how much control you want to place on the people issuing the FWO. If the limit is too high, you will have no control and if it is too low you will be authorizing trivial items. For your own comfort it is probably better to start with a low limit and work your way up to a higher limit. As a guide, for $10 to $15 million dollar projects, $25,000.00 is a reasonable limit for FWOs.

There is no limit to the number of FWOs you can have on a project. Make sure the FWO is closed once the work is completed. On projects it is handy to have one FWO as a slush fund. This is used to catch those small jobs that have to be done but for which it is neither convenient nor efficient to write out a separate FWO. An example of this would be moving furniture in your site office. The cost is minimal, but the work has to be done, therefore it goes through the slush fund. Once the limit is reached, the FWO is closed out and another one started. Just be careful what you charge through this type of FWO as there is a tendency to try to hide things in them.

Audit all unit price and time and material work, i.e., all material quantities and man hours should be verified. The contractor should provide the backup paperwork for his unit price invoices and you should not pay unless you get this information. The unit prices should be negotiated at the award of the contract and as shown in the contract document. Remember that before contract award is the time to challenge the unit prices, not when the first invoice comes in. Usually not enough attention is paid to this point. It is left out of the negotiations and when the first unit price invoice comes in, you wish you had paid more attention during the negotiations. To save a lot of time, it is easier to get a lump sum price to do the extra work but the scope has to be well defined. Otherwise, you will have extras to the extra.

Be aware that some contractors make a living on extras and you will have a rough time if your scopes are not defined. This is how some contractors price their jobs. They look at the amount of the extras and lower their price accordingly, as they know they can make it up on extras. Sometimes you will have to look at the claim to see if it is something that should be understood by the contractor and is a normal part of the work process. If you are not sure talk it over with your supervisor. At other times the mistakes or omissions will be cut and dried. If you made a mistake in the tender documents, live with it and do not take it out on the contractor because he found it or you are afraid to tell your supervisor. This is why detail in your Tender Documents is important.

The claim for an extra should be in writing from the contractor and all paper work you do should cross reference any control numbers the contractor uses. It is imperative that all changes to the contract be identified, field work orders written, and put into the system as soon as possible for cost control.

An FWO should not be used to account for changes in scope. The funds for the FWOs are usually from the project contingency. You do not want your contingency going to scope changes; therefore, changes in scope are covered by a change order (CO). Figure 9.2 is an example of a standard Change Order Request form. Because there is a different authorization process, there is essentially no dollar limit. COs usually need management approval, so you will have to convince management that the change is needed. The Change Order Request form in Figure 9.2 is completed and submitted to management for approval. Sometimes FWOs are collected and when 10 have been written they are totaled and a CO written to cover them all. There are different methods of handling FWOs and COs, so check your company policy to see how they should be handled.

Contract Change Order Request

CO No.: _____

Date: _____   Originator: _____

Contract: _____   Contractor: _____

Bid to Construct: _____   Other Changes: _____

Description of Change (List drawings if applicable.):

_____
_____
_____
_____

Breakdown of Costs (Use Attachments if Necessary)			
System	**Description of Work (List Drawings If Needed)**	**Contractor Purchase**	**Contractor Labor**
	Subtotal		
	Total		

1.   Attach copies of competitive quotes if they are available.
2.   List any exclusions to the price.

Approved by: _____
Project Manager

Figure 9.2 Change Order Request Form

Because a FWO is an extra to the main contract or purchase order, make sure funds are available to cover their cost. The funds in your budget have to be allocated to the purchase order that covers the FWO. This means you will have to revise your purchase order to allocate additional funds to cover the extra work and may have to revise it again if the allocation is exceeded. If the job is small, you can record the extra costs on your Project Cost Report and make one revision when the job is complete, provided you know what all your costs are. In all cases you should have an up-to-date record of your extra costs and the dollar value even if approximate. The above mentioned method of collecting 10 FWOs and issuing a CO to cover them does not take away from the fact that you have to make sure you have money available to cover the FWO. Each FWO has to have funds allocated to it as it is written. As part of the cost reporting system, throughout the project you will have to estimate what you feel the extra costs will be to complete each purchase order.

Back Charges

One of the hardest project costs to control is back charges to vendors and/or contractors. They are usually a source of conflict, especially if the vendor is having trouble and loosing money. For this reason they have to be well documented. The Notice of Back Charge form shown in Figure 9.3 helps provide the required information.

The vendor should be contacted before any work takes place to see how he wants to handle the repairs and subsequent back charge. All details of the discussion should be filled into the form. This should be done in ink, not pencil. You should have an hourly rate for all the plant staff whose work will be backcharged to the vendor. Sometimes the vendor will deal directly with the contractor to have the repairs done and you do not get involved, which is okay. If the vendor is from out of your jurisdiction, your local contractors may not want to deal directly with the vendor and you will then have to be the go-between.

Notice of Back Charge

Purchase Order No. or Contract No.: _____

Company Name: _____

Description of Problem: _____

_____

_____

_____

_____

_____

Name of Individual Contacted: _____

Date and Time Contacted: _____ a.m./p.m.

Contact Made By:

Phone _____ Phone Number:_____

Fax _____ Fax Number:_____

Attach copy of telephone record or fax.

Remedy to Problem: _____

_____

_____

_____

_____

_____

Estimated Cost: $_____   Back Charge FWO No.: _____

Does Vendor Accept Back Charge: Yes ☐   No ☐

Originator & Date   Vendor Acceptance & Date

_____   _____
        signature   signature

_____   _____
        print name   print name

Distribution:

Vendor / Contractor / Project Manager / Accounting / Purchasing

Figure 9.3 Notice of Back Charge Form

A field work order is issued to your contractor to cover the vendor's work and a back charge issued to the vendor to cover the cost of the work. These two documents have to be cross referenced for easy retrieval and control by the accounting group. The back charge amount is deducted from the vendor's invoice. You have to make sure that all the vendor's money is not paid out at the end of the job if there are back charges outstanding. If you have a back charge, accounting has to keep track of the back charge amounts so that the vendor is not overpaid. If you are aware of back charges you should start deducting the money as soon as possible.

If you have back charges, make sure the accounting group is aware of them and that they have a method of tracking them.

Inspection

On small jobs you or your designate will inspect the contractor's work, whereas on larger jobs an inspection company or consultant will be hired. Specialized inspection such as concrete sampling and testing, soil compaction, x-raying, NDT, welding, etc. should be contracted out unless you are trained in it. Some plants have people on staff who are trained in some of these areas and they can be used if desired.

Make daily inspection rounds at different times of the day so the contractor does not know when you will be around. Talk to the people on the tools and listen to what the problems are. You will find out that there are rumors every day as well as people complaining about all kinds of things. Before you challenge the contractor about any issues, make sure you have the facts and the story correct, otherwise you will make a fool of yourself and will lose creditability with the contractor.

In the field, your role as an inspector is a fine line that should not be crossed. You have to remember that the role is inspection and you must not direct the contractor on how to do the work. You can tell him what he can not do if it affects the mill operations, but the minute you tell him how you want him to do the work you are liable for the outcome. His job is to figure out how to do the work within the confines of the mill parameters. So remember it is **INSPECTION** not **DIRECTION**.

The contractors working on the project are in business to make a profit. As an inspector you may not agree with the costs to the plant for some of the work or you may feel that the contractor is making too much money. This is irrelevant to the inspection process, which should not be used to prevent a contractor from making a profit. The time to dispute the costs was during the tendering process. For the plant, the inspection process is to ensure that the plant gets the installation it is paying for and to protect its investment. As an inspector you should be fair.

The detail and thoroughness of your tender documents including the Scope of Work will determine how easy it will be to be fair. The more detail and definition in the documents, the easier it will be to be fair. Sometimes you will have to do your documents in a hurry and there will be a lack of detail; however, as long as you are aware of what is missing, what assumptions you are making, and what your intent was, you can still be fair. You want to have a working relationship with the contractor, otherwise it will be a long, hard project. Remember that there is no "free lunch," even in construction. When you beat up on a contractor at the front end during negotiations to get something for nothing, you will usually pay for it somewhere else during the project.

During the construction phase of the project treat everything the contractor does as if a court case will result. This means you should keep meticulous notes, record phone conversations, publish minutes, keep a daily diary, and do not rely on memory for details. Avoid compromising situations with the contractors who are working for you or who may work for you in the future. If you use common sense, you will be able to keep out of trouble.

Contractor Meetings

If the project is fairly big and there are several contractors, you should have weekly project coordination meetings involving all of the contractors, subcontractors, some plant personnel, and the consultant, if necessary. These meetings should be held at the beginning of the week (Monday mornings) and should be a brief summary from the contractors about where they will be working that week. This lets everyone know what is going to happen and if there are any conflicts, they should be stated. Any items that affect the project as a whole should be discussed here.

For each contract there should be a separate contractor meeting, again at the beginning of the week, and this should cover the contract in detail. These meetings are held at the beginning of the week so that any problems can be taken care of during the week. Holding a meeting later than Wednesday is not very productive. These meetings should be held every week without fail and the following items should be covered:

- Safety:
  Review all injuries for the past week, i.e., first aids (FA), medical aids (MA), and lost time accidents (LTA). Keep a running total of each category and for LTAs describe the details of the accident. Separately discuss any site safety issues.

- Manpower:
  Review the current manpower broken down by trade and manpower planned for the next week. Keep a cumulative total by trade.

- Technical Issues:
  This should cover construction, drawings, and specification problems.

- Schedule:
Review past week's schedule and explain why there was any deviation from it or the master schedule. Discuss the schedule for the coming week.

- Progress Monitoring:
Overall, are the contractors maintaining the schedule? What are the percentages complete for the various areas.

- Extra Work Orders Issued:
List field work orders written or claims against the job for the past week.

- Amendments to Previous Meetings:
List any changes to the last meeting or make changes to the wording the contractor takes exception to.

These meetings should have minutes recorded, typed, and issued, as they will become part of the contract documents. They should be issued as soon as possible after the meeting for the items to be acted upon. They are not much use if issued three or four days after the meeting.

These meetings are important and should not be taken lightly. This is the major form of communication between the contractor and the owner's senior management. It is also the forum for the contractor to have his problems with the owner recorded. These minutes are an official contract document and could be used in future claims or a lawsuit. As you get into the project, you will find the contractor will start to use these meetings to have his project problems recorded. Sometimes you will see yourself being set up for future claims; however, to make the process work, you have to record what transpires, even if you are in the wrong. If you don't have regular meetings, the only recourse for the contractor is a paper war, which you will loose.

At the first meeting you have with the contractor tell him that when reviewing extra work orders you want to know:

- What extra work orders were issued in the past week;
- Are there any outstanding extra work orders from the past week, whether valid or in dispute, that he intends to submit?

If he does not tell you of these outstanding extra work orders, advise him that you will not consider them in the future. What you are trying to do is keep on top of all the claims against the contract. Try to resolve extra work claims as soon as possible. If you can't, you may have to leave them to the end of the job but make sure you have good documentation on the problem.

Diary

Keep a daily diary of what has transpired during the day. Entries into the diary should be made throughout the day, as you will forget important items if you try to sit down at the end of the day and fill it in. It is important to keep a daily diary. You have to make up project reports and this diary is the source of what has transpired. Also, the information you put in your daily diary will be used in the case of future claims and lawsuits. The diary should be written as if it will be used in a court case, which means concise notes. To see how well you write diaries and memos, look at some things you wrote two or three months ago and see if you understand what the items were about.

The plant may have standard diary forms (see Figure 9.4); however, the daily diary should include the following:

- Date:
  The date should be written out, i.e., 10 May 2001. Because date conventions differ, you do not want confusion as to the correct date. August 7, 2001 can be written as 08/07/01 or 07/08/01.

- Weather:
  This is important for future claims, as you may want to know if it was raining, cold, etc.

- Description of Work Performed:
  Keep track of important items worked on, achievement of milestones, etc.

- Critical Incidents:
  Such as accidents, safety issues, labor problems, down time and length, etc.

- Material Deliveries:
  Note important deliveries, especially if the delivery had problems. The stores will have their records but you should make a note of the events that affect you.

- Major Equipment on Site:
  This refers to cranes, size and number, etc. Your contractor should be providing this under a separate document.

- Manpower on Site:
  This should be provided by the contractor on a separate form.

When filling in your diary, write it with the thought that if the contractor files a claim two months from now, do you have enough information to refute it? This cannot be emphasized enough.

Daily Project Diary

Project: _____

Engineer: _____ Date: _____

Weather: _____

Work done today, hindrances or delays, instructions or conversations
with contractors, material received, visitors, FWOs, CCOs, fire safety,
evacuations:

_____
_____
_____
_____
_____
_____
_____
_____
_____
_____
_____

Safety: _____

_____
_____
_____
_____
_____

Delays/Problems: _____

_____
_____
_____
_____
_____

Figure 9.4 Typical Daily Project Diary Form

Photos

If necessary, progress photographs should be taken at set intervals to document progress. Additional photos should be taken, as required, of anything that will be buried or covered up, the inside of equipment that has been opened for inspection, heavy lifts that are out of the ordinary, etc.

With site pictures, you will find out two things:

1. The picture will come in handy for the oddest reason and you will be glad it was taken;
2. The picture will only contain part of what you are interested in. However, part of a picture is better than no picture at all.

The pictures should be labeled and filed away for safekeeping. Control who borrows them and where the pictures are taken to when removed from the office.

Case History 17

During the construction phase of one project we were looking for a very large wood crate full of parts required for installation. We searched the plant site but could not find anything matching the description. We were going to challenge the vendor about not delivering it when someone produced a photo taken during the offloading of equipment and to one side was the crate in question. It was not the main subject of the picture, but fortunately it was in the picture. The crate was about 10' W x 10' L x 20' H and had a big X on it, so it should have been easy to find. We searched the site again but never found it. To keep the job moving, we had to reorder all the parts that were in the crate. I worked in the plant for a year after startup and never saw the crate or its contents around the plant.

Do not let others take pictures around your plant site. Your plant should have a policy about photography and you should enforce it. This would apply to union business agents, vendors, contractors, and the media. You do not want pictures taken and then showing up in vendors' advertising for all your competitors to see. Let management decide what picture they want to allow.

Correspondence

Any correspondence from the contractor should be acted upon and responded to as soon as possible. It has to be done in a timely fashion and failure to respond implies that you agree with what the contractor has written. This could be detrimental if a legal claim is made against the project. Do not procrastinate about responding to matters you do not feel comfortable about. If you need time to gather information, send back a note stating that is what you are doing and that you take exception to what was written. When you get correspondence of this nature you will be thankful you kept such good notes in your diary.

Always make every effort to get the contractor all the information and material he needs to complete the work. The late delivery of drawings and/or material can lead to claims by the contractor for extras. This is why you have to keep track of all correspondence that goes out of your office. Drawings should be listed on a transmittal with the revision number and issue number. All documents, including drawings, that you receive should have a "RECEIVED" date stamped on it. You will find the contractor will also date stamp the drawings when he receives them just for the purpose of tracking drawings and claims.

If the contractor files a claim stating that late delivery of material or drawings held up certain work, you will have to go over his schedule and manpower to see if he was really held up or if he could have been doing something else. The contractor may not have any choice with small jobs, but on large jobs with many functions the contractor can usually move his manpower around.

Deficiency List

Throughout the construction phase you should keep a running deficiency list (sometimes refereed to a "punch list") of the deficiencies known to you at that particular point. The items should be brought to the contractor's attention if the deficiency item affects the current work. It is important to bring up the items you feel are deficiencies as it may be a difference in the interpretation of the contract that has lead to the deficiency. This allows the problem to be solved now and not at a later date. The deficiency list will be updated throughout the construction and will become part of the contract documents at the end of the job. As the work nears completion, an official deficiency list can be prepared (see Figure 9.5). Do not give any deficiency list to a contractor containing items that are not part of his scope, or items that are part of the normal construction process and will get completed as the work finishes.

At the point of substantial completion, a deficiency list of items is given to the contractor and final completion is not achieved until the deficiencies are satisfactorily addressed. You will have to sign off the deficiency list for the final completion to be reached and the final money paid out.

Deficiency List

Contract No.: _____ Date: _____

Issue No.: _____ Revision No.: _____

Contractor: _____ Sheet: _____ of _____

Area: _____ Issued By: _____

Structural ☐   Electrical ☐   Instruments ☐
Mechanical ☐   Piping ☐   Miscellaneous ☐

Please return to: _____ of _____
when items are complete.

Item No.	Description	To Be Completed By	Date Completed

Figure 9.5 Deficiency List

Equipment Checkout

As the equipment installation progresses, keep track of where each piece of equipment is in its installation cycle. On large projects this is done using "tags" and an Equipment Record of Installation form (Figure 9.6). The equipment, electrical, and instrumentation tags are all different colors, waterproof, require a special pen for marking, and when all the items have been signed off, the equipment is ready for testing. At this point a green tag is placed on the equipment and the owner accepts it. The Equipment Record of Installation form has the same information as the tag.

The theory of the tag system is to place a tag on every piece of equipment to identify it with its name and equipment number. The tags represent different stages in the installation of a piece of equipment. As each stage is completed, you and the contractor sign off on the tag indicating that the stage is completed and the equipment is ready for the next stage. This way you can go up to a piece of equipment and see what stage of the installation process it is at. However, the tags are not secure and can be destroyed by various means. It is not uncommon to get half way through the installation and have tags go missing. For this reason, if you are using tags, you should use the Equipment Record of Installation form in conjunction with the tags. The Equipment Record of Installation form is the official document of what stage the installation is at. When you witness a stage, you and the contractor sign off the tag and the Equipment Record of Installation form at the same time. Most plants do not have tags and it is not worth the expense to make them up for small numbers of equipment. Therefore, you should use the Equipment Record of Installation form by itself. Keep these in a three-ring binder for easy access.

The are similar forms for the electrical and instrumentation checkout (Figures 9.7 and 9.8).

Equipment Record

Equipment No.: _____   Engineer: _____

Drive No.: _____   Motor No.: _____

Description: _____

Remarks: _____

_____

		Completed			
		Contractor		Engineer	
Stage	**Description**	**Date**	**Sign**	**Date**	**Sign**
1	Set in Place				
2	Leveling of Primary Equipment & Baseplate				
3	Grouting				
4	Attachments to Equipment				
5	Alignment				
6	Lubrication				
7	Electrical Checkout				
8	Equipment Guards				
9	Test Run				
10	Instrumentation Checkout				
11	Alignment Recheck				
12	Dowels Required **Qty.** **Size**				
	Equipment				
	Drive				
	Motor				
	Lugs Required **Qty.** **Size**				
	Equipment				
	Drive				
	Motor				
13	Installation Complete				

Figure 9.6 Equipment Record of Installation

Electrical Equipment Checkout Record

Equipment No.: _____   Engineer: _____
Drive No.: _____   Motor No.: _____
Description: _____
_____

Stage	Description	Completed			
		Contractor		Engineer	
		Date	Sign	Date	Sign
1	Equipment in Place & Installation Ready for Test				
2	Insulation Resistance Test D.C. Hi-Pot ____ KV ____ Res. Test Feeders _____ Ohms Test Controls _____ Ohms				
3	Equipment Checkout				
4	Test Run Amps, Under Load ____ Amps, Without Load ____ Volts, Under Load ____ Volts, Without Load ____				
5	Installation Complete				

Figure 9.7 Electrical Equipment Checkout Record Form

Instrument Checkout Record

Instrument Tag No.: _____ Engineer: _____
Instrument Type: _____
Description: _____
_____

| Stage | Description | Completed | | | |
| | | Contractor | | Engineer | |
		Date	Sign	Date	Sign
1	Receive & Inspect				
2	Shop Calibrate:				
	Calibrate By: _____				
	Range: _____				
3	Installation Complete Ready for Checkout				
4	Instrument Ready For Functional Checkout				
5	Instrument Operational				

Figure 9.8 Instrument Checkout Record Form

Substantial Completion and Final Completion

There are two points that define the completion of the construction work—substantial completion and final completion. Your plant should have a policy on when these two points are reached.

In general, *substantial completion* is reached when the work has been completed to a point that the work can be used for its intended purpose. The work does not have to be complete, but you have to be able to use what has been installed in a safe manner that will not damage the plant or equipment. At this point, you have to have a deficiency list that you and the contractor agree upon. The deficiency list has a cost assigned to it reflecting what it will cost to complete the work. If you agree to substantial completion, a legal document is filled out giving the date substantial completion was achieved (see Figure 9.9). These Certificates of Substantial Completion are standard forms for each jurisdiction and can be purchased at your local stationary store. The contractor can now apply for partial release of his holdback. If you have been following procedures, you have been holding back 10 – 15% from every invoice the contractor has submitted. Partial holdback will be the holdback minus twice the value of the work left to complete on the deficiency list. Before you can pay out partial holdback, the contractor has to prove to you that all his subcontractors and workers have been paid, there are no claims against the job, and his worker's compensation is up to date.

Certificate of Substantial Completion

Contract No.: _____ Date of Inspection: _____

Contractor: _____

Contract Title: _____

This certificate refers and relates to the contractual agreement dated _____ between MHS Engineering Services Inc. (hereinafter referred to as the owner) and the contractor.

In accordance with the contract between the owner and the contractor, the owner hereby certifies that on the basis of an inspection jointly carried out, the work is at this date, namely _____ (hereinafter referred to as the "Agreed Date of Substantial Completion"), suitable for the purpose for which it was designed.

The agreed date of Substantial Completion shall be regarded as the date of Substantial Completion for all purposes whatsoever and, without restricting the generality of the foregoing, for determining the rights, duties and obligations of the owner and the contractor under the agreement between them, as well as for all purposes under the Builder's Lien Act.

A Deficiency List of items to be completed or corrected is attached hereto. The contractor undertakes to complete or correct the work listed as quickly as possible in accordance with the terms and conditions of contract taking into consideration the availability of materials and labor. The estimated date for completion of deficiencies is

_____

Value of deficiencies as shown on attached list is $_____

The holdback against the contractor is to be released in accordance with the contract document, less an amount equal to twice the agreed value of the deficiencies, which amount is to be released seven (7) days after all the deficiencies are completed.

The guarantee of the contractor with regard to accepted portions of the work is to commence on the agreed date of Substantial Completion.

Except as in this Certificate expressly provided, this Certificate does not affect any rights, duties, or obligations between the owner and the contractor.

MHS Engineering Services Inc.	Contractor
Signed By _____	Signed By _____
Title: _____	Title: _____
Date: _____	Date: _____

Figure 9.9 Certificate of Substantial Completion

Final Completion is reached when the contractor has completed all items on the deficiency list. If during the operation of the new equipment, other deficiencies appear, these can be added to the deficiency list. Again, if you agree to final completion, a legal document is completed indicating when final completion was achieved (see Figure 9.10). These are standard forms that can be purchased at your local stationary store. Your jurisdiction defines the time period the contractor has to wait before he can apply to you for the release of the holdback. This can range from 30 to 45 days. Before payment, the contractor has to provide proof of payment to subcontractors and workers, that there are no claims against the plant, and that worker's compensation payments are up to date.

For large projects there can be a substantial amount of money in the holdback account. On a $10 million dollar project this can amount to $1.0 to $1.5 million. This is money the contractor has to carry the financing on, so there is a desire on his part to get the holdback paid out as soon as possible. You will get requests from the contractor to substitute other methods of protecting the plant and pay out the holdback. One of these is to have the contractor's bank issue a letter of credit in the plant's name for the value of the deficiency list. If process guarantees are involved, the amount of the letter of credit would have to be substantial. Your purchasing group and plant management would have to agree to this letter of credit, as it is a major change.

Certificate of Final Completion

Contract No.: _____ Contractor: _____
Contract Title: _____

As the owner, we hereby certify that the works of the above contract have been satisfactorily executed as of _____ under the terms and conditions of the contract and certify the following:

Value of awarded contract. _____

Total of approved extras. _____

Total due under contract and
approved extras. _____

Deduct for work that will not be completed under
the scope of the contract. _____

Final total amount certified by the owner as
being due under the contract. _____

Payments up to and including progress payment. _____
Dated _____

Claims submitted by the contractor and not
certified by the owner. _____

This certificate is issued pursuant to clauses in the above mentioned contract.

MHS Engineering Services Inc.   Contractor
Signed By_____   Signed By_____
       Title:_____   Title:_____
       Date:_____   Date:_____

Figure 9.10 Certificate of Final Completion

Commissioning and Startup

Although most people use the terms interchangeably, commissioning and startup are two distinct phases. *Commissioning* is running all the equipment in automatic control without any feedstock. Once this has been completed, startup can begin. *Startup* is the running of all the equipment in automatic mode with feedstock in order to produce product. The turnover sequence chart in Chapter 10 clearly defines the two stages. Mechanical completion usually occurs at the end of commissioning and is the end of your contractor's contract. This means that startup has to be funded from a different budget item. Contractors will supply you with tradesmen for startup but this is on a time and material basis. On most projects these stages are a blur and there is confusion on the part of the field staff. These stages should be defined at the beginning of the project so everyone involved understands them.

Commissioning and startup of the equipment have to be considered at the beginning of the project to make sure:

- there is money in the budget for startup;
- equipment is constructed in the required sequence;
- any special equipment or material required for commissioning is available;
- special tradespeople are available;
- the necessary numbers of people required are available.

Vendors, if they are busy, may have trouble supplying the appropriate people to meet your schedule. There may be other plant shutdowns in the area limiting the number of tradespeople you can get to work on your project. To keep all those concerned informed about your project, commissioning and startup has to appear in your project schedule.

Commissioning and startup should be part of the design process. The project should be broken down into systems, e.g., water, air, steam, conveyors, pumps, etc. and constructed in the order of commissioning and startup. On very large projects with a lot of equipment, the commissioning should be scheduled and tied into the overall construction schedule. This breakdown can be further refined as construction progresses based on problems that have developed. Be aware that deviating from the commissioning schedule can delay the overall startup. For smaller projects, the commissioning and startup is a one- or two-day affair that requires organizing the appropriate team members and doing everything over this time period. On large projects this process can take a month or more and could be done after hours, on weekends, etc. —whatever is required.

For the startup phase, the vendor process guarantees should be known and available to all team members so the correct operating procedure can be followed to ensure the guarantees are met. Getting the adequate running time on new equipment can be a problem in operating plants, but every effort should be made to get the equipment and process guarantees satisfied and out of the way.

The next chapter contains a commissioning procedure you can use. It should be reviewed in detail to ensure that it is suitable for your jurisdiction. In the procedure is a turnover sequence that shows what work has to be completed during the commissioning and startup phases.

Training

Prior to the commissioning phase the employees should be trained on the new equipment. This should not be left until the very end. The operators should be trained and ready to assist in the commissioning by operating the equipment for the vendors. The training should include both operations and maintenance personnel. These are usually different training sessions, as you do not necessarily want the operators included in a maintenance training session. Sometimes maintenance personnel will go to the operators training session to better understand the maintenance required on the equipment.

Depending on the equipment, the training could be:

- A short classroom session of up to several hours combined with a field visit to the equipment for an explanation of its features;
- A long classroom session of several days combined with several visits to the equipment for explanation of its features;
- Hands-on training on the floor at the equipment;
- Sending operators and maintenance personnel to another plant for several weeks or months;
- Spread over a month with training at different stages of equipment installation.

The methods used for training can be different. The trainer could use:

- Written manuals
- Videos
- Slides
- Overheads
- Models
- Computer Simulation
- Actual Installed Equipment or "On The Job Training"
- Hands-on such as fire extinguisher and fire fighting exercises

The person doing the training could be:

- The salesman who sold the equipment
- The equipment erector
- The startup person
- Someone brought in specifically to do the training

Some problems you can run into with training include:

- Trainer has no presentation skills.
- Trainer speaks poor English.
- Employees do not understand English very well.
- Employees' experience is not up to par for the job and the training has to be aimed at a lower level than normal.

Plant Project Engineering Guidebook

- Trainer is too technical for what is needed.
- Trainer is disorganized at training session.
- Trainer is not adequately prepared for the group size.
- Trainer does not know his material.
- Training material is not adequate.
- Proper teaching aids for the training are not supplied by the plant.
- Poor facilities for the training sessions.
- Scheduling and priority issues between training and other plant activities.
- Trainer gets off the subject and nobody brings him back.

Unless you are doing on-the-job training, the employee training should take place before any equipment is started up by plant personnel. Some of the plant operators may know more than the trainer; however, to protect the plant's interests those operators have to take the training as well.

To ensure that any training is carried out effectively, certain information must be given to the vendor about the conditions he can expect, and you must receive certain information from the vendor so that you can plan accordingly.

Following is the body of a letter that has been used to exchange information with vendors.

General Planning Information

The following information has been provided to ensure that the instructors are well informed and prepared to provide effective training.

Number of Trainees

Unless otherwise instructed, you will be responsible for training up to 15 plant employees. The employees will be a mix of maintenance technicians, lead operators, operators, and department managers.

Employee Schedule:

8:00 to 12:00	Employees In
12:00 to 1:00	Lunch Break
1:00 to 5:00	Employees In

Training Location

The training room will be located in a small office in the plant. Due to the number of employees involved and the size and shape of the room, information should to be presented by overhead projector, TV/VCR, and/or handout so that all employees can see the data.

Scheduled Training Times

Since we are currently coordinating multiple tasks and training, it is critical to our overall productivity to start and stop training as scheduled. Once training is set, if the date or time must change, please let us know as far in advance as possible to reset and reassign the work schedule. If other arrangements are not made we will assume the training will be on the date and time previously specified and will ensure the employees are on time and prepared.

Training Agenda

To allow us to better prepare for the training sessions, please fax a training agenda as far in advance of the scheduled training date as possible.

Equipment Required

To allow us to better prepare for the training session, along with the agenda, please list what type of equipment you will need for the training session:

TV/VCR	_____
Overhead Projector	_____
Specialized Equipment (List what type)	_____

Plant Project Engineering Guidebook

Contact Names

Please provide the name and title of the training provider.

The following individuals can assist you if further information is required:

> John Doe
> Phone:
> Fax:

You can change any of the above to suit your situation.

You may have trouble getting some of the more advanced presentation systems if you are in a remote location. Computer projectors are hard to find and expensive to rent. Be prepared to discuss some of the presentation issues with the vendor.

Once the training date is set and your people organized it becomes difficult to rearrange schedules. A few days before the training session confirm with the trainer that he will be there and confirm what equipment you are providing.

For the training session itself, make sure you or your designate attend the training session. You want to know what went on at the training session and you want to hear it from an unbiased source. You will be surprised at the comments you will get.

Case History 18

{I was coordinating the training on a large project and was sitting in on some of the training sessions. One session was actually quite good and the plant operators thought it was the best training session they had had to date. A month later when they were complaining about the training they were getting, they stated they had not received training from the above vendor! It did not take long to turn that complaint around. Had I not been there, I would have had to take their word for it and rescheduled another training session at my cost.}

At the end of the training session, have the trainees evaluate the trainer. Some questions you can ask are:

1. Was instructor prepared?
2. Did the instructor know his material?
3. Was the training session organized?
4. Was the written material adequate?
5. Was the visual material adequate?
6. Did the training session meet the stated objectives?
7. Were there any communication/language problems?
8. Was the material easy to understand?
9. How would you rate this training session?

These questions should be rated on a scale of 1 to 5 as follows:

1 = Very Satisfied
2 = Satisfied
3 = Neither Satisfied or Dissatisfied
4 = Dissatisfied
5 = Very Dissatisfied

Allow space at the bottom of the page for any comments.

A few days after each training session you may want to test the employees on what they have learned.

As a project engineer, the extent of your involvement may be only organization; however, what is outlined above is what you can expect to happen.

Chapter

10

COMMISSIONING PROCEDURES

Introduction

Commissioning procedures are not very common in most plants. Those that are available are probably from the original plant startup and can be too cumbersome for small plant projects. The following commissioning procedure has been modified so you can use it for your small plant projects. It is written based on the assumption that you have hired an outside consultant to do the engineering and construction management. You can modify and use individual sections of this procedure as they suit your situation. The mechanical, electrical, and instrumentation sections are detailed in what commissioning is and how it is to be done. If you assemble Chapters 10, 11, 12, and 13 into one document, you will have a complete commissioning procedure. Even if you are not commissioning equipment, the information presented is something you should know for future use.

For this commissioning procedure the overall flowsheet is broken into different groupings based on what equipment can be run without interfering with the other equipment down the line. The groupings were color coded and numbered on the master flowsheet. A list is made up with the numbers of all the equipment in each group. From this flowsheet and equipment number list everyone involved in the startup will know exactly what the commissioning sequence will be and what equipment has to be completed in order to commission a group.

The Turnover Sequence Chart, under item 9, shows the different phases involved in the commissioning and startup. When you look at the phases you will see they are clearly defined as to what takes place during each phase. It also shows when mechanical completion, substantial completion, and final completion occur. This document should be discussed with the contractor at the pre-award meetings so there is no misunderstanding of the phases and when they end. At this meeting, you should get a feel for the manpower requirements and length of time needed to carry out the phases. This should help in your project estimating at the front end of the project.

The following administrative, mechanical, electrical, and instrumentation procedures are used for the commissioning (engineering checkout and system testing) activities for the plant equipment and systems. Manufacturers' representative's information and manufacturers' technical manual instructions supplied with the equipment are used throughout the commissioning and supplement these procedures.

Engineering checkout is performed during and after construction is complete to verify that the equipment and systems are capable of functioning in accordance with design and manufacturers' specifications. When all equipment checkouts and tests are satisfactorily completed, the checkout activities continue in logical stages until individual systems are commissioned. Once commissioned the entire process is placed in operation and raw material is run through the normal flow paths using the Human Machine Interface (HMI) to monitor and control the operating functions. Commissioning is complete when all equipment and systems have been verified as satisfactory and documented using the following procedures and supplemental procedures, as applicable, contained in manufacturers' technical manuals. Mechanical completion is certified when checkout is complete. This is also substantial completion.

After mechanical completion, the system testing phase begins with the water trials, which is the running of all the equipment in automatic mode with water instead of process liquids or raw material.

Representatives from the major equipment suppliers will participate in the startup phase to ensure that their equipment operates properly. Part way through the system testing, final completion will be reached. This occurs after the deficiency list from substantial completion is completed.

Operations personnel will be integrated and involved in the systems testing phase. The classroom portion of the training will include in-mill periods to review layout and other details of the equipment, but the majority of hands-on training will be accomplished during systems testing. Maintenance training will also be part of the systems testing phase.

In order for commissioning and startup to be carried out smoothly and safely the events have to be well organized and responsibilities clearly defined. As you will be running equipment that may not have it's guards in place it is imperative that people are aware of what is going on and when. The following administrative procedure outlines definitions, the groups involved, responsibilities of all the people involved, activities involved in the phases, and how the different systems are turned over to the next group. The procedure will also describe the lock out and tagout of equipment, and commissioning schedule issues.

Administrative Procedures

AP-1   Startup Program Administrative Procedure
AP-2   Lockout/Tagout Procedure
       (Control of Hazardous Energy)
AP-3   System Boundary Identification Procedure
AP-4   Schedule

Administrative Procedure AP-1
Startup Program

1.0   Purpose and Scope

1.1   To define the phases of the commissioning (Engineering Checkout and System Testing) and assign responsibilities.

1.2   To provide a method to ensure that all equipment and/or systems have been proved operational before the system is turned over to the Plant.

2.0   Definitions

2.1   Construction Inspections and Checks

Those inspections and checks performed by Construction to ensure that the required construction activities are completed prior to commissioning and startup.

2.2   Phases of the Commissioning and Startup Program

The Plant Startup Sequence is divided into three phases listed below and shown graphically in AP-1, Attachment 1:

- Construction Completion Phase
- Commissioning
    - Engineering Checkout
    - System Testing Phase
- Startup Phase

2.3   Commissioning

Commissioning is defined as the performance of those inspections, system tests, and trials, required to ensure that a portion of the plant is ready for first time start-up and continuous operation. It is the dry running of the equipment systems, without introducing any raw materials, to verify the proper functioning of the equipment system and its associated control system. Test materials may be used.

2.4   Startup

Startup is the running of all the equipment in automatic mode with feedstock in order to produce product. for the purpose of commencing commercial operations.

2.5   Mechanical Completion

Mechanical completion will occur when all construction work is complete and the work is ready for system testing, although some work may still be needed. The work left to be done, however, will allow the Plant (or system) to be operated without damage to the Plant (or system).

2.6   Deficiency List (Punch List)

There are two Deficiency Lists that you will have to prepare.

1. Construction Deficiency List
2. Commissioning Deficiency List

The Construction Deficiency List defines the work that remains to be done for the contractor to complete the construction. The project engineer prepares this list.

The Commissioning Deficiency List is prepared by the project engineer but with input from the operations group. This list addresses design issues that affect the operation of the plant.

2.7   Functional Circuit Check

A verification of the integrity of control circuits by simulation or manipulation of every contact and/or device within the control circuit with control power applied.

2.8   Instrument Calibration

Individual adjustment of instruments and control devices utilizing predetermined values and verifying acceptable quantitative accuracy obtained.

2.9   Functional Loop Check

A set of tests performed to verify the functional integrity of an instrumentation control loop. These loop checks are performed with all loop components installed, electronic loop energized, and/or pneumatic loop pressurized. This includes a loop check from the primary device to the HMI screen and from the HMI screen back to the final device.

3.0   Participating Groups

3.1   Plant

Plant representatives located at the job site.

3.2   MHS Engineering Services Inc. (MHS)

MHS Project Manager located at the job site.

3.3   Contractor

A subcontractor to MHS involved in the construction of the Plant.

4.0   Responsibilities

4.1   The Plant

The Plant is responsible for the following activities:

4.1.1   Participation in the system turnover procedure as outlined in this Administrative Procedure.

4.1.2   Operation of all permanent plant equipment to support the commissioning schedule under the direction of the MHS representative.

4.2   MHS Engineering Services Inc.

MHS shall be responsible for the following:

4.2.1   Furnishing all of the engineering documents and information necessary for the completion of construction.

4.2.2   Furnishing the engineering documents required to complete the commissioning and startup phases.

4.2.3   Furnishing engineers on site during the commissioning and startup phases to provide an interface on design and engineering problems, as required.

4.2.4   Providing the technical training of operators.

4.2.5   Completion of all the systems to support the commissioning and startup schedule.

4.2.6   Participation in the system turnover procedures as outlined in this Administrative Procedure.

4.2.7   Inspection activities performed by the construction supervisors and engineers in accordance with the erection and installation specification(s).

4.2.8   The Construction Manager shall be aware of the startup priorities and how construction interfaces with the commissioning and startup schedule.

4.2.9   Assuring plant equipment has been lubricated and maintained in accordance with the vendor instruction manuals and that all applicable records are on file.

4.2.10   Assuring that the preliminary and final coupling alignments are completed in accordance with the equipment vendor instructions and that all applicable records are on file.

4.3   MHS Commissioning and Startup Organization

4.3.1   Startup Engineering
Startup Engineering will develop all the procedures required for the system test and startup phases prior to equipment checkout and testing.

4.3.2   Startup Manager
The Startup Manager is responsible for the overall direction and conduct of the commissioning and startup program for plant equipment and systems. He has responsibility for scheduling and directing the efforts of those assigned to him in the performance of testing and checkout activities. He shall coordinate the interface activities of the startup, engineering, construction, and the Plant operating personnel required to accomplish the commissioning and startup program.

4.3.3   MHS Startup Engineers
The Startup Engineers assigned to the site are responsible to the Startup Manager for performing assigned checkout, system testing, and startup program activities.

The Startup Engineers shall:

- Conduct the test procedures required for the performance of all phases of the System Test and Startup Program.
- Review and approve all data obtained to ensure that systems and equipment are performing in accordance with engineering and design specifications.
- Prepare systems for turnover to the Plant operators and assist and direct them during the system testing and unit startup phase of the startup program.

5.0   Commissioning & Startup Activities

5.1   Construction Completion Phase

During the construction of the plant systems, construction supervisors and engineers shall perform necessary and required inspection to ensure that completed installations are in accordance with the latest engineering and design information and specifications.

5.1.1   Electrical:
All electrical wiring, (power, control, and instrumentation) shall be installed and terminated in accordance with design documents.

5.1.2   Mechanical:
All mechanical equipment shall be installed in accordance with design documents. Final coupling alignment shall be done after Startup Engineers have run the associated motor satisfactorily. First fill lubrication shall be done during this construction phase. The appropriate lubrication and alignment forms (not attached) shall be used to record the applicable data. Any rotating equipment operational problems associated with coupling misalignment or pipe stress is the responsibility of the construction group to correct.

5.1.3   Piping:
        All piping shall be installed in accordance with design documents. The construction group shall conduct required piping system hydrostatic tests. Any piping changes required to obtain a correct coupling alignment shall be completed by construction. All piping, tanks, tops of tanks and chests shall be cleaned and flushed thoroughly prior to being turned over to the startup group.

5.1.4   Instrumentation:
        All instrumentation devices and piping shall be installed and calibrated in accordance with design documents.

5.2   Commissioning Phase

The MHS Startup Manager and his staff shall direct and supervise the checkout and system testing personnel during this phase.

All system checkouts, system testing of different sections of the plant, flushing, and operational testing shall be done during this phase by the MHS startup group with electricians, pipefitters, and technicians provided by construction. The Plant operators may be involved in this phase.

5.2.1   Electrical:
Conduct electrical function circuit checkout of all electrical controls. An authorized contractor or an electrical startup engineer shall check out all electrical systems. The following tests shall be conducted:

5.2.1.1   Insulation resistance test of electrical equipment to be done prior to energization (power cables).

5.2.1.2   Checkout and verification of control circuitry. Electrical schematics shall be highlighted to indicate checkout is completed.

5.2.1.3   Test and checkout of medium and low voltage switchgear, motor control centers (MCC), and molded case circuit breakers.

5.2.1.4   Checkout of AC electric motors.

5.2.1.5   Checkout of motor operated valves, dampers, and gates.

5.2.2   Mechanical - Piping:
All mechanical equipment shall be inspected, checked out, and flushed by the startup personnel.

5.2.3   Instrumentation:
All instrument calibration and testing shall be done during this phase. Functional loop checkout includes checking each loop from the field device to the HMI computer or final control device. The startup group will do this. Final instrument calibration, where required, will be done during the startup phase.

The startup group instrument test technician will do the following:

5.2.3.1	Verify that all instrument devices including factory installed skid mounted instruments are calibrated.
5.2.3.2	Conduct functional instrument loop checkout on all instrument loops.
5.2.3.3	Troubleshoot instrumentation problems during startup and operation.

5.3   Startup Phase

The MHS Startup Engineers shall accomplish the startup phase using the manufacturer's Startup Engineers and the Plant operators. This phase includes all tests and checks, which are necessary to ensure satisfactory commercial operation. This phase shall include:

5.3.1   Placing all systems in automatic operation and obtaining ready for running program status.

The Plant startup phase shall be directed by the MHS Startup Manager and coordinated with the Plant Manager.

5.4   General

5.4.1   Craft labor will be provided by Construction and directed by the Startup Manager in support of checkout and startup activities.

5.5   Vendor Services Management - Startup

The Startup Manager will have overall responsibility for determining the need for, and requesting all vendor service representatives for, startup and training. Startup personnel will sign Manufacturer's Representatives Daily Reports that include actual hours worked. Copies of these reports will be sent daily to the Startup Manager. If warranty work or work of disputed responsibility is performed, it should be so noted on this report. When a vendor is preparing to leave the site, startup personnel shall notify the MHS Startup Manager for discussion of outstanding items, (i.e. final trip report, material, or information needed from vendor, items that may be returned prior to payment). Contractors who require vendor support will coordinate their own vendor activities.

5.5.1   General:
         All startup and test engineers will schedule vendor representatives in a timely and efficient manner to ensure their hours are kept to a minimum and are within the allocated budget.

5.5.2   All vendor representatives shall be required to prepare daily work descriptions and time records. The vendor's form must describe the vendor's daily activities in detail and indicate to whose account the time is charged.

6.0   System Identification

MHSs Engineer shall prepare a "System Boundary Identification List" identifying each plant system with a system number. Each system shall be subdivided into subsystems for further identification. This list shall be used to identify and describe system boundaries for the turnover program.

All system correspondence, technical data, and information shall be filed using the System Boundary Identification List as an index. A System Turnover Number shall identify all systems turned over from Construction to Startup.

7.0   System Turnover Method

The following method shall be used to document and turn over a completed system (Turnover Package) from Construction to Startup and then from Startup to Plant Operations (see AP-1, Attachment 1) for System Turnover Sequence.

7.1   Construction Turnover for Engineering Checkout Form

The MHS Construction Manager (CM) is responsible for initiating the Construction Turnover Form. The CM shall verify the contractor's work is complete before signing the Construction Turnover Form.

The CM shall coordinate all construction Punch List items and list them on Form Number 100A which shall be attached to Turnover Form 100 (AP-1, Attachment 2). The completed Turnover Form (100), including the System Punch List (100A) shall be forwarded to the Startup Manager.

The Startup Manager or his designate shall perform an inspection of the system, adding to the System Punch List items that are either not completed or need corrective action to comply with design and shall indicate to the CM any items that must be completed prior to acceptance of the system.

When the Startup Manager is satisfied that the system described on the Turnover Form is ready for system testing, he should sign the Turnover Form.

The CM shall maintain the System Punch List on a system basis during the remaining system test and construction program. The System Punch List shall be kept up to date so that it reflects the status of the system. Additional items will be added to the punch list during the system test and startup program as it becomes necessary. The CM shall ensure that uncompleted and unsatisfactory items are resolved and removed from the Punch List on a timely basis to support the test and startup program. The items on the Punch List shall only be construction related. A separate Design/Operations Punch List will be maintained by the Startup Manager for design and operation related issues.

The CM shall report the status of each System Punch List at the Startup Coordination Meetings.

When the Startup Manager has accepted a system, the System Startup Engineer shall conduct a thorough inspection of the system. He shall also conduct all system testing utilizing manufacturer's representatives and Plant operators.

The Startup Manager or his designate shall determine when the system is ready to be turned over to the Plant operators. The Startup Manager shall initiate the Turnover Form 100 describing the boundaries and exceptions. The Turnover Form shall be forwarded to the Plant Project Engineer.

The Plant Project Engineer or his designate shall perform an inspection of the system, adding to the Design/Operations Punch List items that require action to comply with design and shall indicate to the Startup Manager any items that must be complete prior to acceptance of the system.

When the Plant Project Engineer is satisfied that the system or subsystem described on the Turnover Form is satisfactory for further testing and operation, he shall sign the Turnover Form.

7.2   Final System Turnover Package for Startup.

The *System Turnover Package* shall consist of the following documents and shall be forwarded to the Plant Project Engineer at the time of system turnover for startup. The MHS Startup Group will compile and issue the turnover packages to the Plant Project Engineer.

The system turnover package shall consist of the following:

Section 1 Turnover Forms
>   A Construction Turnover Forms for Engineering Checkout - Form 100

Section 2 System Boundary Identification
>   A System Boundary Identification Drawings (Process & Instrumentation Drawings - P&ID's).

Section 3 Mechanical Data
>   A Alignment record forms
>   B Vibration records
>   C Mechanical test reports and recorded data

Section 4 Electrical Data
>   A Electrical test data forms as applicable from electrical test procedures with test data recorded
>   B Electrical schematic drawings highlighted indicating checkout complete
>   C Vendor electrical test data (if applicable)

Section 5 I&C Data

    A Input/output lists highlighted indicating checkout complete

    B Instrument device calibration sheets

Section 6 Vendor Information

    A Vendor information useful for system operation/maintenance such as vendor trip reports or technical information

7.3   General

7.3.1   Routine maintenance, lubrication, calibration, and administrative control shall become the responsibility of the Plant Operations Group after they have accepted equipment/systems from the MHS Startup Group and the equipment/system is mechanically complete.

8.0   Technical Data Documentation

The following documentation shall be used to verify proper equipment preparation and operation:

8.1   Alignment Record Form

This form shall be completed by the Contractor on all rotating equipment at the time of coupling alignment. This form shall be used to verify coupling alignment before piping is attached to a pump and to verify the correct alignment reading after piping is attached to a pump. The MHS CM is responsible for this form.

8.2   Vibration Record Form

The Vibration Contractor will supply this form and it will be used to record equipment vibration at the time of initial operation. Vibration data shall be taken on uncoupled motors as well as on coupled motors and equipment. This form is the responsibility of the MHS Startup Manager.

8.3   Electrical Data Forms

All data will be completed by the electrical test contractor responsible for system electrical checkout and testing. All nameplate and test data shall be recorded, including motor rotation as outlined in the applicable electrical test procedure.

9.0   Attachments

1. System Turnover Sequence
2. System Turnover Form
3. System Punch List

AP-1, Attachment 1
MHS Engineering Services Inc. - Turnover Sequence

Installation Phase	Commissioning			Startup	
Construction Completion	**Checkout**	**System Testing**		**Integrated Operation**	
Equipment installed.	Bump & run motors.	Demonstration of system electrical control functions.		Demonstration of furnish addition and production of final product.	
Grouting complete.	Set limit switches.	Demonstration of system equipment functions.		Functional demonstration of individual equipment's efficiency and operating parameters.	
Mechanical alignment.	Flush & clean equipment using installed equipment.	Operator on-the-job training.		Functional testing and adjustments to major equipment.	
Hangers & restraints installed.	Steam blow.	Check pipe hangers/supports for acceptable restraint when at design temperatures.		Optimization of chemicals to produce the required quantity and quality of product.	
Hydrostatic testing complete.	Final alignment.	Dry run systems without material.		Final optimization prior to performance test.	
Initial equipment lubrication.	Verify relay & overload settings.	Continue to tune instrument loops.		Performance test.	
Pipe clean & inspected.	Functional checkout of components & circuits.				
Equipment cleaned & inspected.	Thermal oil hydrostatic test & boilout.				
Mechanical "as built" drawings provided.	Stroke valves.			Start of warranty	
Electrical cable installation checks.	Check vibration of rotating equipment.			Normal operation	
Wiring & termination checks.					
Continuity tests completed.					
MCCs installed/meggered.					
Electrical "as-built" drawings provided.					
Construction Completion	Substantial Completion & Mechanical Completion	Contract Deficiencies		Final Completion	

AP-1, Attachment 2 - Form 100
Construction Turnover for Commissioning and Startup

System Name: _____

Turnover Package No.:   _____   Date:_____

Construction work, testing, and inspection of the system or portion thereof contained in the above Turnover Package is complete and is hereby released to MHS Startup Team for Commissioning and subsequent operation.

Mechanical work complete and inspected: _____

Electrical work complete and inspected: _____

Instrument work complete and inspected: _____

Construction Manager: _____   Date:

Startup Manager: _____   Date:_____

Startup Turnover to Plant Operations

Commissioning and initial operation of the above system is complete and the system is hereby released to Plant Operations for Startup and normal operation. Punch List items have been cleared with the following exceptions:

_____

_____

Startup Manager: _____   Date:_____

The above system is hereby accepted by Plant Operations, who will operate and maintain this equipment under the direction of the Startup Manager or designee through the completion of the Performance Test.

Project Engineer: _____   Date:_____

AP-1, Attachment 3
System Punch List

System No:				Turnover Package No:		Sheet: _____ of _____
Item No.	Responsible Group	Date Added To List	Scheduled Comp. Date	Actual Comp. Date	Deficiency Description	

signature _____   CM _____   date _____

Plant Project Engineering Guidebook

Administrative Procedure AP-2
Lockout Procedure

1.0   Purpose

1.0   To assign responsibilities to ensure a safe lock out.

2.0   To provide a method so equipment and machinery are locked out in a safe manner.

2.0   Definitions

Lock out	means the application of a lock, or several locks, to the control devices providing the primary source of energy to the machinery or equipment. The source of energy may be electrical, air, gas, or hydraulic. In electrical systems, control devices means disconnect switches on MCC's or local disconnects. **They do NOT include control buttons or control circuits.**
Lock Out Tag	a standard red and white **DO NOT OPERATE** tag with a blank space for writing the required information on it. Plastic laminate tags are preferred.
Lock Out Log	a hard covered notebook kept at the lock out board. The log book is to record the date, lock number, equipment placed on, time, name of person placing the lock, and the reason for the lock placement.
Lock Out Board	a yellow board with black lettering that holds locks, keys, hasps, and the Lock Out Log

3.0 Application

Employees, contractors, and/or staff working on or in close proximity to equipment, which has the potential for unexpected operation, movement, release of energy or release of hazardous materials, must have the source of that energy tagged and whenever possible locked to prevent the release of that energy.

A part of a machine, piece of equipment, device or thing shall be cleaned, oiled, adjusted, repaired, or have maintenance work performed on it only when,

a) motion that may endanger a worker has stopped, and

b) any part that has been stopped and that may subsequently move and endanger a worker has been blocked to prevent its movement.

Where the starting of a machine, piece of equipment, devise or thing may endanger the safety of a worker,

a) control switches or other control mechanisms shall be locked out, and

b) the effective precautions necessary to prevent such starting shall be taken.

This procedure deals primarily with locking out electrical energy sources, but shall also be applied to other equipment that fits the above criteria such as; dryer hydraulic drive.

4.0 Lock Out Boards

Lock out boards will be located in the following areas:

a) adjacent to and on the north side of the press

b) on column W13 beside the sanders

c) inside door 15 on the east side

5.0   Responsibilities

5.1   Regular Maintenance and Breakdowns

Locks are required if the machine requires daily maintenance or a breakdown requires work on the equipment.

Lock #1 Operator
Lock #2 Maintenance Personnel
Lock #3 Outside contractor, if necessary

5.2   Preventative Maintenance (Scheduled)

Locks are required if the machine requires Preventive Scheduled Maintenance.

Lock #1 Operator
Lock #2 Maintenance Personnel
Lock #3 Outside contractor, if necessary

Maintenance personnel must inform the Operator when the job is complete.

5.3   Shut Down Due to Unsafe Conditions

Lock #1 Operator
Lock #2 Plant Manager
Lock #3 Maintenance Personnel
Lock #4 Outside contractor, if necessary

The Plant Manager must be informed, by the operator, the reason for the lock out. The Plant Manager will place his lock on the machine.

5.4　Equipment Not in Use

Equipment that is not in use will have one lock placed on the main power, by the Operator. There must be a tag attached to identify that the equipment is *not in use*. Keys for this equipment will be kept in the Plant Managers office.

5.5　Responsibilities of Employees

Each worker who works upon the machinery or equipment requiring lock out procedures shall be responsible for:

a)　Placing his locks on the control devices
b)　After locks have been applied, the affected machinery or equipment shall be checked by a designated employee, who has applied a lock, to ensure that it cannot be operated by pushing the start button, if electrically operated, or opening valves if air or hydraulically operated.

NOTE: Locking out upstream pressure in air or hydraulic lines also locks in downstream pressure, which could cause an unwanted cycling of the machine. Sudden release of this pressure can be dangerous. Therefore, it is recommended that the residual pressure be drained off slowly.

a)　Removal of his own locks on the completion of the work.

5.6   Contractors

Contractors will be responsible for placing their own locks on the equipment before work commences. The operator will make locks available to the contractor. Contractors must return the locks to the Operator when the equipment is safe to operate. The logbook must be filled out in accordance with Section 7 of this procedure.

5.7   Lock Out Tag

One Lock Out Tag must be affixed to the hasp, clearly identifying the name of the key holder, reason for the lock out, and the date.

5.8   Control of Keys

Employees shall not leave their keys in their lock when it has been applied to a control device.

5.9   Multiple Locks on One Control Device

When more than one lock is required on one control device a scissors clip shall be used. The last hole on the scissors should be left empty of a lock so that an additional scissors clip can be used if necessary.

6.0   Verifying Isolation of Power

After hasp and locks are in place, verify energy source has been shut down by pressing the start button.

7.0   Log Book

Attached at each Lock Out Station will be a Lock Out Log Book. The Operators will be responsible to ensure the Lock Out Log Book is filled out every time a lock is placed on the equipment.

8.0   Change of Control Procedure

If the work required, cannot be completed in the same shift, the keys for the Lock Out must be given to the next shift's Operator. The Operator will distribute the keys to the appropriate personnel. Each person receiving a key will immediately confirm by physically checking that each key controls one different lock on the lock out. Each person will familiarize himself or herself with the reason(s) for the lock out.

9. 0   Removal of Locks

Every person, who placed a lock on, must take off his or her own lock. Before a lock is cut off, the Manager or Operator shall first make every effort to contact the individual who put the lock on, and they must be present to ensure that the machinery or equipment starts up safely.

The second set of keys is in the key cabinet located in the Manager's office.

10.0   Safe Start-Up

Once people are clear of all moving parts, electrical systems safe, all work is complete, tools are stored properly (will not damage machine) all guarding is in place and the operator is satisfied that the equipment is safe to run, only then will the equipment be started up.

11.0   Training

All appropriate personnel must be trained in this procedure.

Employee/Contractor Acknowledgement

I _____ acknowledge receipt of a copy of MHS Engineering Services Inc, Lock Out Procedures. I have read these and understand them. I have been given an opportunity to discuss these with my supervisor who clarified any concerns I have.

Signed _____

Company _____

Date _____

Supervisor Acknowledgement

I _____ have issued the above employee/contractor with a copy of MHS Engineering Services Inc., Lock Out Procedure. I have discussed these with the employee/contractor who has demonstrated his understanding of them.

Signed _____

Date _____

Note: It is a requirement of MHS Engineering Services Inc. that employees complete this acknowledgement within seven days of commencing work.

12.0   Quick Reference

Quick Reference – Lock Out Procedure

The following are steps to take when the equipment you are operating becomes unsafe or is not operating properly.

a)   Shut down all sources of energy to the equipment, (main power). Do not attempt to adjust or repair the equipment.

b)   Have a coworker stay at the equipment to ensure against accidental start up.

c)   Place a lock on the main power to equipment. Test equipment to ensure power has been isolated.

d)   Report lock out and reason for lock out, to your manager.

e)   Operator will ensure the Lock Out Log Book is filled out. Including dates, names, equipment, and reason for lock out.

f)   Maintenance personnel will place a third lock on equipment prior to work.

g)   When work is completed all personnel must remove their own locks and return locks to the lock out board

Administrative Procedure AP-3
System Turnover Boundary Identification Procedures

1.   System Index

This form contains:
- System Number - These numbers are listed in perceived order of startup.
- Description of Boundary - description of area the system number refers to. The areas can be broken down into subsystems if required to facilitate system testing.

System Number	Description of Boundary
100	Raw Material Preparation
200	Raw Material Washing
300	Grinding
400	Drying

2.   Turnover Package Index

The following table contains:
- Turnover Package Number - Note that there are gaps left between numbers when there is more than one turnover package within the same system. An example is System 100 which has 105A & 110A as Turnover Package Numbers.
- P&IDs that apply to the turnover package. To produce these P&IDs, your flowsheet must be in electronic form. Using the boundaries you have set up, take each system off the master flowsheet and put it on to an 11 x 17 sheet of paper. This way you have an individual flowsheet for each turnover package and less confusion.
- Description of Boundary - description of area the P&ID refers to.

Turnover Package Number		
Turnover Number	**P&ID**	**Description of Boundary**
105A	FR-100A	Primary Raw Material Preparation
110A	FR-100A	Secondary Raw Material
205A	FR-104A	Raw Material Washing
305A	FR-105A, 105B	Deflaker Through to Cyclone
405A	FR-107A	Dryers

3.   **System Boundary Drawings**

P&ID's are marked with the turnover package boundaries. P&ID's and electrical one-line drawings are not included in this document.

The following table contains:
- Number – the Turnover Package Number;
- Description Boundary – description of the area the equipment is located in;
- Equipment Numbers – equipment numbers that are in the boundary area and that are included in the turnover package.

Turnover Package/Equipment Numbers Index		
Number	**Description of Boundary**	**Equipment Numbers**
105A	Primary Raw Material Preparation	101, 102, 103, 104, 105, 106
110A	Secondary Raw Material	107, 108, 109, 110, 111, 112,113, 114, 115,
205A	Raw Material Washing	120, 201, 202, 203, 211, 212, 213, 214, 215,
305A	Deflaker Through to Startup Cyclone	225, 226, 227, 228, 229, 230, 231, 232, 233,
405A	Drying	301, 302, 303, 304, 305, 306, 307, 308, 309

Administrative Procedure AP-4
Commissioning Schedule

1.0   Purpose

The purpose of the Commissioning Schedule is to establish dates when systems are to be turned over for Commissioning and Startup.

2.0   Scope

The system will be turned over for Commissioning on the dates shown on the schedule. Prior to system turnover, the system shall be inspected by the responsible construction and startup supervisors to assure that all of the details are complete and the system is ready for Commissioning and Startup. Documentation shall be in accordance with Administrative Procedure AP-1.

Guideline 155

{This is a schedule you develop for your project. Do not make the commissioning part of the overall construction schedule. Your construction schedule should only show a date when commissioning starts. Keep commissioning and startup as separate schedules and in greater detail than the construction schedule.

At this point you have the administrative portion of the commissioning procedure set up, and everyone involved should now know the order of startup and what constitutes startup. This should allow the construction group to know what to focus on to complete the work on schedule.}

Chapter

11

MECHANICAL COMMISSIONING
PROCEDURES

This section of the commissioning procedure presents the methods to be used to carry out the checkout and startup of mechanical equipment. Review your drawings and standards, then revise this section to reflect what you are trying to accomplish. Also, make sure these procedures match your plant procedures and the codes you are using.

Mechanical Test Procedures

MTP-1 System Inspection Procedure

MTP-2 Initial Operation of Rotating Equipment

MTP-3 Piping System Cleaning Procedure

MTP-4 Air System Cleaning Procedure

MTP-5 Pressure Testing Installed Piping Procedure

Mechanical Test Procedure MTP-1
System Inspection Procedure

1.0   Purpose and Scope

1.1   To establish a procedure to inspect a system or a portion of a system prior to checkout and system testing.

1.2   To establish a method which MHS Construction and MHS Startup Engineers will use to inspect systems and report any deficient items found so appropriate corrective action may be taken.

2.0   Prerequisite

2.1   The MHS Construction Manager (CM) and MHS Startup Manager (SM) will schedule the inspection (walkdown) based on the startup schedule and discussions during startup meetings.

3.0   Procedure

3.1   Inspection

> 3.1.1   The CM and SM will inspect each system and associated equipment that is to be turned over.
>
> 3.1.2   Inspection will include a walkdown of the system and the listing of items not complete at the time of turnover. Attachment 1 will be used as a guideline during system walkdown and inspection. Form 100A, will be used to record all punch list items.
>
> 3.1.3   If the CM and SM find the system satisfactory, that is, either complete in all aspects or only deficient in items, which will not prevent system testing in a safe manner, the SM will sign the Turnover Form and take care and custody of the equipment/system. If the equipment/system is found unsatisfactory, the Construction Turnover Sheet and the System Punch List will be returned to the CM for corrective action.

3.2   System Punch List

3.2.1   A System Punch List, Form 100A will be prepared for each system.

3.2.2   The Master System Punch List and the resolution of the items will be kept current by the CM.

3.2.3   The CM will monitor and update the System Punch List as often as necessary to ensure completion of items.

4.0   Start-Up Forms

The forms attached to this section shall be used to record mechanical startup inspection data. After the forms have been completed, they will be included in the System Turnover Package in accordance with AP-1.

5.0   Attachments

5.1.1   System Inspection and Checklist

MTP-1, Attachment 1
System Inspection and Checklist Guidelines

The following list should be used as a guideline to inspect systems and equipment within a turnover boundary:

A. Mechanical Inspection and Checks

1. Piping installed according to flow diagrams _____
2. Pipe supports installed (slippers installed and lubricated) _____
3. Pipe hangers installed in locations shown on drawings _____
4. Spring hangers cold set _____
5. Pipe restraints installed _____
6. Pipe snubbers installed and adjusted _____
7. Thermal insulation installed (if applicable) _____
8. Heat tracing installed according to flow diagrams _____
9. Personnel protection installed for hot pipes (not designed for thermal insulation) _____
10. Equipment properly grouted _____
11. Equipment properly mounted (equipment to base plate, base plate to concrete) _____
12. Equipment and area clean _____
13. Equipment properly lubricated: lube sheet available and completed _____
14. Air Operated Valves (AOV) and Motor Operated Valves (MOV) that have special shipping packing have been repacked with proper packing _____
15. All grease connections of valve stem yoke bushings, equipment bearings, glands, plug valves, etc. have been lubricated _____
16. Temporary piping installed, if required, for initial operations _____
17. Temporary strainers screens installed with applicable Differential Pressure gages, if required, for initial operations _____
18. Local vents and drains with pipe caps as appropriate are installed. There may be valves added due to layout consideration that may not be included on the flow diagram _____
19. Valve locking devices provided according to flow diagram _____

Plant Project Engineering Guidebook

A. Mechanical Inspection and Checks (Continued)

20. Startup consumables on site (e.g., filter cartridges) _____
21. Coupling guards provided _____
22. Air operated valves installed, including air supply, filter/regulator and regulated air pressure indication _____
23. Relief valve code inspector seals intact, if set points are not to be re-verified on site _____
24. Piping expansion joint shipping stays removed. If applicable, tie rods installed and adjusted. If tie rods are required for hydrotests, the rods should have been re-adjusted for operation or removed _____
25. Coupling alignment completed and alignment sheet reviewed _____
26. Manual valve packing installed and packing gland tight _____
27. Hydrostatic testing complete, if applicable _____
28. Pneumatic testing Complete _____
29. Piping clean and inspected _____
30. Valves installed and identified _____
31. Equipment identified and tagged _____
32. "As-built" drawings provided _____
33. Technical manuals available and complete _____

B. Electrical Inspections and Checks

1. Power cables pulled, terminated _____
2. Motor operated valve motors oriented in proper position _____
3. Control cables pulled, terminated _____
4. Motors and switchgear properly grounded _____
5. Continuity and megger test completed _____
6. Switchgear bus bars bolts torqued _____
7. MCCs installed/meggered _____
8. Annunciator installed _____
9. MCCs installed and connected _____
10. Ground straps installed _____
11. Protective relaying installed _____
12. Electrical "as-built" drawings complete _____
13. CT circuits verified to ensure no open circuits _____

C. Instrumentation and Control Inspections

1. Solenoid operated valves and their associated controls installed _____

2. Field and panel mounted instruments installed, including all flow, temperature, vibration, pressure, level differential pressure, and associated wiring _____

3. Pressure and level sensing tubing installed, including applicable heat tracing, instrument isolation valves, vent valves, drain connections, pressure calibration connections, and applicable condensate pots with fill connections for level sensing lines _____

4. Flow transmitter sensing tubing slope in proper direction and without low point dips. Water and steam lines slope down to transmitter and draft sensing lines slope up to transmitter _____

5. All instruments identified with mark number tag _____

6. All instrument electrical wiring complete between instrument devices and PLC/HMI _____

7. All instrument device calibration records available _____

8. Calibration "sticker" on instruments indicating calibration _____

Reviewed/Approved: _____ Date: _____
                                        MHS CM

Reviewed/Approved: _____ Date: _____
                                        MHS SM

Mechanical Test Procedure MTP-2
Initial Operation of Rotating Equipment

1.0   Purpose

1.1   The purpose of this procedure is to establish the methods for:

    1.1.1   Initial operation of rotating equipment.

    1.2.2   Documentation of initial operating characteristics of the equipment.

2.0   Prerequisites

2.1   The following prerequisites must be completed before initially attempting to operate a piece of equipment.

    2.1.1   Personnel involved shall be thoroughly familiar with the applicable vendor manual and drawings, placing special emphasis on:

        2.1.1.1   Installation requirements such as: mounting, lubrication, and alignment.

        2.1.1.2   Design limitations

        2.1.1.3   Pump sealing configurations and requirements including packing adjustment and seal venting.

        2.1.1.4   Run-in requirements.

        2.1.1.5   Restart limitations for large motors.

    2.1.2   Nameplate data information.

    2.1.3   Electrical controls have been checked out and tested in accordance with the applicable Electrical Test Procedure (ETP) and documented on the proper form.

    2.1.4   Personnel shall be thoroughly familiar with the system and equipment files associated with the piece of equipment to be tested to assure they are cognizant of all available information.

    2.1.5   Personnel shall be thoroughly familiar with P&IDs and actual as-built configurations of the equipment and systems.

2.1.6 Lubrication has been verified and completed lubrication data sheet is on file.

2.1.7 Final alignment data has been verified and the alignment data sheet is on file.

2.1.8 Instrumentation has been calibrated and loop checked to monitor performance of the equipment, or enough calibrated test equipment installed to verify expected parameters.

2.1.9 The uncoupled and coupled run of all motors will be the responsibility of the appropriate startup engineer.

3.0 Precautions

3.1 The following precautions must be followed as a minimum to ensure protection of personnel:

3.1.1 Lockout/Tagout shall be established in accordance with MHS procedures to provide adequate flow boundaries and electrical protection.

3.1.2 If the startup engineer deems it necessary, the area around the equipment to be operated initially will be roped off and only authorized personnel will be permitted in this area.

3.1.3 Equipment and flow path have been visually inspected immediately prior to initial operation to verify system boundaries and integrity.

3.1.4 All prerequisites must be complete and personnel cognizant of required and expected parameters such as minimum suction head, discharge shut-off head, expected operating discharge pressure and flow, no load and full load motor current and direction of proper rotation.

4.0 Procedure for Initial Uncoupled Motor Run

4.1 Megger readings have been completed prior to initial run.

4.2 All prerequisites pertaining to motor run have been completed.

4.3   All support equipment such as instrumentation and control power are energized and in service.

4.4   Check control switch in off position and pull-to-lock position if pull-to-lock feature is provided.

4.5   Rotate motor by hand to ensure it is free to rotate.

4.6   Where necessary, a phase rotation meter shall be used to verify proper motor rotation prior to energizing the motor.

4.7   Rack in the main motor breaker.

4.8   Momentarily energize the motor (bump) then return control switch to off. Verify rotation is in the proper direction.

   Note: When starting large motors it is preferable to let starting current decay before stopping the motor in order to minimize stresses applied to main breaker. The exception to this is a bearing arrangement that could be damaged by reverse rotation.

4.9   Energize the motor and record current readings and motor rotation in accordance with the appropriate electrical test procedure.

4.10   The motor will be run until temperatures stabilize (about 15 minutes). Motors 400 hp or larger will be run for a longer period of time as determined by the startup engineer. Stator temperature will be monitored when stator RTDs are provided.

CAUTION

If any run-in limits specified in the vendor manual are exceeded, immediately stop the motor.

Note: The following data must be closely monitored and documented during the run-in.

- Motor current, all phases
- Bearing temperature
- Stator temperature
- Vibration

4.11   After electrical test procedure data sheets have been completed for uncoupled run, verify motor run is acceptable according to design parameters.

5.0   Procedure for Initial Coupled Run

5.1   Complete valve line-up necessary to establish flow path, except pump discharge valve which should be positioned as dictated by impeller flow characteristics.

5.2   Pump suction is filled, vented, and the discharge lines are filled and vented as much as possible.

5.3   Place in service the external lubrication system, if provided.

5.4   Place in service the bearing cooling system, if provided.

5.5   With breaker tagged and racked out, ensure motor and driver equipment can be rotated by hand.

5.6   Ensure coupling guard is installed and secure.

5.7   Place in service the packing gland sealing and lubricating system, if provided.

5.8   Use plant operators to operate all coupled permanent plant equipment.

5.9   Start the drive unit and, if applicable, begin to open the discharge valve very slowly, observing suction pressure (to ensure it remains adequate) and discharge pressure. When discharge pressure no longer decreases as the discharge valve is opened, the system is full and the discharge valve is now to be opened fully.

5.10   The equipment being initially operated will be run until temperatures (bearing, winding seal, etc.) stabilize. Equipment with a drive motor larger than 400 hp will be run for about 2 hours or as determined by the startup engineer in charge.

Note: The following data must be closely monitored during the initial coupled run.

- Motor bearing temperature
- Pump bearing temperature
- Motor stator temperature
- Packing gland or seal temperature
- Motor current

CAUTION

If, during the initial run-in, any limitations set forth by the vendor are exceeded, immediate corrective action should be taken.

5.11   Complete all electrical and mechanical data sheets for coupled run of equipment.

6.0   **Acceptance**

6.1   Initial operation is satisfactory as determined by review of the test data and considering the conditions the equipment was operated under.

7.0   Procurement of Vibration Data

7.1   Purpose

    7.1.1   To establish the method of monitoring vibration.

7.2   Precautions

    7.2.1   When using probe pickup, use caution in the placement of the probe to prevent entanglement of the probe in the rotating element.

    7.2.2   Do not "drape" the cord across the rotating element of the equipment being tested.

    7.2.3   Do not wear loose clothing in the immediate vicinity of rotating equipment.

7.3   Prerequisites

    7.3.1   Equipment must be running as close to normal configuration as possible.

7.4   Procedure

    7.4.1   Vibration monitoring:
    Horizontal equipment will be monitored in 3 planes—horizontal, vertical, and axial. Vertical equipment must be monitored in two planes—vertical and horizontal. All vibration monitoring will be performed using the velocity RMS setting on the meter.

7.5   Acceptance

    7.5.1   Vibration levels are considered acceptable if they are less than the maximum level specified by the vendor. If vendor information is not provided, the startup engineer will determine if the vibration level is acceptable using vibration analysis charts shown in the attachments.

Mechanical Test Procedure MTP-3
Piping System Cleaning Procedure

1.0   Purpose

To establish the method for conducting water flushes that clean the system and minimize danger to personnel and equipment.

2.0   References

2.1   Applicable vendor instruction manuals

3.0   Precautions

3.1   A valve lineup must be completed prior to initiating the flush to assure the required flow path has been established.

3.2   Boundary tags have been installed as necessary to protect personnel and equipment.

3.3   When open-ended flushes are being conducted, the area of the flush discharge must be roped off and personnel not involved in the flush must stay clear.

> **Guideline 156**
>
> {The flushing is easier to do after hours when there are fewer people wandering around the plant. The piping is usually up high, there is a lot of water, and there can be a lot of refuse coming out of the pipe at a high velocity. This has to be well planned.}

3.4   When open-ended flushing is being conducted, protect any electrical equipment that may be wetted by the flush discharge.

4.0   Prerequisites

4.1   Initial operation of the equipment necessary to support the flush must have been satisfactorily performed before the flush commences.

4.2   A marked up print (P&ID) showing the flow path or paths has been prepared.

4.3   Involved personnel have been briefed on the flow path, equipment limitations, and the requirements for acceptance.

4.4   Temporary flushing pumps and equipment must be sized to obtain required flow rate.

5.0   Procedure

5.1   The system line-up has been completed according to the attached marked up print.

5.2   All boundary tags have been installed, according to the attached marked up print.

5.3   Temporary strainers have been installed in the lines according to the marked up prints, as applicable.

5.4   Flush the closed loop system utilizing the permanent installed pump or temporary flushing pumps as approved by the engineer. Suction strainers shall have differential pressure gauges to allow monitoring of the strainer. The strainer shall be cleaned initially within one hour of start of the flush, then as required to maintain the specified differential pressure.

5.5   Open-ended flushing cleanliness will be determined by placing a sample funnel with a suitable strainer/filter in the discharge stream. Closed loop flushing cleanliness will be determined by inspection of strainer/filters installed in piping system.

6.0   Acceptance

6.1   See MTP-3, Attachment 1 - Acceptance Criteria and Approval

6.2   The responsible system startup engineer shall determine when the system is clean based on the vendor's cleanliness requirement.

7.0   Restoration

7.1   All temporary strainers/filters have been removed. All piping has been restored to normal configuration.

8.0   Attachments

8.1   Acceptance Criteria and Approval

MTP-3, Attachment 1
Acceptance Criteria and Approval

8.1   Acceptance Criteria and Approval

The applicable system vendor instruction manual has been reviewed
and the following acceptance criteria established:

Guideline 157

{There are a lot of systems for which there are no vendor criteria. In
these cases you have to flush until the startup engineer is satisfied
that the lines are clean.}

8.2   Visual inspection of strainers or filters shows no significant
         dirt or debris on the strainer or filter.

8.3   The _____ system has been flushed in accordance with
         an approved procedure including all vendor instruction
         manual requirements and found to be satisfactory.

Performed by: _____ Date: _____

Approved: _____ Date: _____
                          MHS Startup

Mechanical Test Procedure MTP-4
Air System Cleaning Procedure

1.0   Purpose and Scope

This procedure defines the method to be used for cleaning the instrument and service air piping, prior to placing it in operation.

2.0   Scope

2.1   Air will be blown down from each branch line to remove construction debris.

2.2   Blow media will be from the air compressors.

3.0   References

3.1   Air system P&ID

4.0   Special Test Equipment

4.1   Test Pressure Source:
- Station air compressors

4.2   Ear Protection:
- Ear plugs
- Ear protectors

4.3   Eye Protection:
- Safety glasses
- Plastic face shield

4.4   Hand Protection:
- Heavy leather gloves

5.0   Prerequisites

5.1   The air compressors have been checked out and put into service.

5.2   The air receivers must have been inspected and cleaned prior to being pressurized.

5.3   The instrument air dryer must be put into service and the entire instrument air header system pressurized.

5.4   All piping and equipment pressure testing has been satisfactorily completed.

6.0   Initial Conditions

6.1   The system is pressurized.

6.2   The air compressors and dryers are operating normally.

7.0   Precautions and Limitations

7.1   Always wear protective equipment for eyes, ears, and hands.

7.2   If airflow exceeds 500 scfm, fluidization of desiccant in dryers may occur, causing filters to plug.

7.3   Ensure that the area around the discharge of any valve being blown down is clear of personnel.

8.0   Procedure

8.1   Open the blow control valve at the end of the instrument air header for 30 seconds.

8.2   Close valve and install cloth bag over valve discharge.

8.3   Open valve for 30 seconds.

8.4   Close valve and remove cloth bag.

8.5   If no significant amount of debris is evident, proceed to next valve.

8.6   If a significant amount of debris is collected, repeat steps 8.1 through 8.6.

8.7   Proceed through all instrument air root valves as outlined above. The airline from the root valve to the instrument will be blown down just prior to placing the instrument in service.

8.8   Repeat the procedure for the service air system by blowing from the service air outlets throughout the plant.

9.0   Acceptance Criteria

9.1   All instrument air root valves shall be blown until no significant amount of debris is visible on the cloth bag.

9.2   The responsible startup engineer shall determine when the system is acceptable based on cleanliness of cloth bags.

10.0   Attachments

1   Approval Sheet

MTP-4, Attachment 1
Approval Sheet

1.   The following person conducted the air system cleaning procedure which satisfactorily cleaned the air header:

Name: _____ Date: _____

2.   Comments:

_____
_____
_____
_____
_____
_____
_____
_____
_____
_____
_____
_____
_____
_____
_____
_____
_____
_____
_____
_____
_____
_____

Approved: _____ Date: _____
                     MHS Startup

Mechanical Test Procedure MTP-5
Pressure Testing Installed Piping

1.0   Purpose and Scope

1.0   This procedure defines the technical and administrative requirements for pressure testing installed piping systems or portions of a system in accordance with the applicable design code. Any conflicts between this procedure and the specification or other controlling document shall be referred to the MHS Startup Manager for resolution.

2.0   General Test Method

2.1   General

    2.1.1   ANSI B31.1 and Others
        Piping designed, fabricated, and erected under this code shall receive a hydrostatic or pneumatic test prior to initial operation. Systems not specified to receive these tests will receive an Initial Service Leak Test (ISLT).

    2.1.2   National Fire Code (NFC)
        All piping classified as Fire Protection System Piping shall be hydrostatically tested in accordance with applicable NFPA codes.

    2.1.3   The pressure test shall include the instrument lines up to the last isolation valve before the instrument. If the instrument lines are not installed at the time of test, an ISLT is acceptable.

    2.1.4   The pressure test shall include all sample lines up to the last isolation valve at the sample rack. If sample lines are not installed, the test shall be up to the sample line connection root valve and an ISLT shall be performed on the sample lines when installed.

Plant Project Engineering Guidebook

2.1.5   The contractor is to provide the testing forms that collect the required information to meet the noted standards.

2.2   Test Pressures

Test pressures for system piping will be in accordance with the code requirements for the boiler and fire protection system. All other systems will be subject to an ISLT at initial operation.

2.3   Definitions

2.3.1   Piping System:
A configuration of material and components which, when installed, provides a flow path for a fluid.

2.3.2   As-Built:
The documentation used to describe what was actually installed.

2.3.3   Inspection:
The physical examination of a process or product to determine its adherence to applicable requirements.

2.3.4   Lay-up:
The protection of piping or equipment after it has been pressure tested to prevent corrosion of interior surfaces prior to subsequent operation.

2.3.5   Pressure Test Supervisor:
Contractor's supervisor responsible for pressure test.

2.4   Test Media

2.4.1   Hydrostatic Testing - If the system cleanliness has been
         established prior to hydrostatic testing, water of a purity
         equal to or greater than that specified in the system
         operating procedure shall be used. If hydrostatic testing
         is performed prior to establishing the final pre-
         operational cleanliness class, filtered water or water
         meeting normal system water quality may be used. At
         no time shall the test media decrease the system
         cleanliness established at the time of the test.

2.4.2   Pneumatic Testing - Pneumatic leak tests shall
         normally be performed with clean, oil-free dry
         instrument air. Clean, dry nitrogen is also
         permissible. In no case shall other than a non-
         flammable gas be used for a pneumatic test.

Note: A test medium other than those specified above may be
permitted when specifically allowed by an approved procedure.

Guideline 158

{You have to a make sure the test media will not contaminate
the final product. An example is thermal oil. If the intended
final product is thermal oil, the lines must be tested using
thermal oil as water will rust the interior of the pipes and
contaminate the oil. The system is also flushed using oil and
by the use of in-line strainers the pipes are cleaned during the
boil out. For these instances of special test materials you have
to be prepared for the possibility of failing the test. If you
have a leak at a weld, what do you do with all the test
material? Make sure you have enough tank storage capacity
to handle the volume.

If using a test medium other than water, it is prudent for the
contractor to pretest the piping with air at a low pressure.

This will determine any leaks before putting the actual test medium in the piping.}

2.5   Schedule

2.5.1   Before any testing can take place, the contractor will submit, for review and approval, a schedule outlining the date, time, and duration of each test.

Guideline 159

{When pressure testing vessels, try to schedule the tests so that you fill and test the largest vessel first, then transfer that water to the next largest, and so forth down the line. This saves time, energy, and water by only getting a large volume of water once.}

3.0   References

3.1   AP-1, Administrative Procedure for the Performance of the Test and Startup Program.

4.0   Special Test Equipment

4.1   Test Pressure Source - The hydrostatic or pneumatic test procedure source shall be capable of producing and maintaining the required test pressures. There shall be provisions for relieving pressure located in the immediate reach of the pressure source operator.

4.2   Test Gauges

4.2.1   Only calibrated test gauges shall be used for pressure testing.

4.2.2   The pressure gauges used for pressure testing shall preferably have a dial graduated over a range approximately double the intended maximum test pressure. In no case shall the gauge range be less than 1 ½ nor more than 4 times the maximum pressure.

5.0   Prerequisites

5.1   Verify from the latest applicable Pressure Test Diagram the hydrostatic test pressure for piping to be tested.

Guideline 160

{The pressure test diagram is a drawing produced by the design engineer showing the test pressures required for the different lines according to the code the line was designed under.}

5.2   A copy of the latest system flow diagram marked up to "as-built" conditions, showing the applicable test boundaries, shall be part of each system pressure test. These flow diagrams shall be marked up and submitted along with the test schedule as required in Section 2.5.1.

5.3   Prior to testing a system, an inspection shall be made to ensure that any drain plugs installed on diaphragm valves are removed to allow checking for leaking diaphragms.

5.4   Prior to testing a heat exchanger, vents and drains on the side not under test shall be opened to allow checks for internal tube leakage.

6.0   Initial Conditions

6.1   Test pressure source, gauges, and overpressure protection are installed and communications established.

6.2   All joints, including welds, branch connections, and regions of high stress, such as regions around openings (weldolets, sockolets, elbowlets, etc.; thickness transitions, reducers, reducing elbows, reducing tees, etc.), flanges, valve bonnets, etc. shall be left uninsulated and exposed for inspection during the pressure test when possible.

6.3 Equipment, such as system relief valves that are not to be subjected to the pressure test shall be either disconnected, gagged, or isolated by a blank flange or similar means. Valves may be used for equipment isolation if the valve is suitable for the proposed test pressure, as determined by the Pressure Test Supervisor. Isolated portions of the system shall be vented to prevent overpressurization.

Guideline 161

{Do not wait until the last minute to decide on what blind flanges to use. There may be a case where a special flange has to be made up for the testing. These should be made up well in advance of the testing and the blind flanges should be kept in a secure location until needed.}

6.4 Pipe hangers have been adjusted or pinned and expansion joints restrained for the test as required.

Guideline 162

{You can run into severe problems if you do not restrain expansion joints. You have to prevent the expansion joint from expanding, which will damage it beyond repair.}

6.5 The official test pressure gauge, for hydrostatic tests only, should be located at the highest point in the system, or elevation corrections shall be made. These corrections are necessary to ensure the required test pressure is applied to all points within the test boundary.

7.0   Precautions and Limitations

7.1   For safety reasons, work in areas of the test shall be limited during the test, as determined necessary by the contractor's Pressure Test Supervisor.

Guideline 163

{This may require testing after hours or on weekends when fewer people are around. Because of the high pressures you are obtaining, large areas may have to be roped off to keep people out. Pressure vessels that are constructed on site have to be tested on site. If the size of the equipment being tested allows, you may be able to cover it with blasting mats to prevent material escaping if the vessel fails.}

7.2   Suitable isolation and venting shall be provided at all equipment, pumps, relief valves, meters, and gauges not tested.

7.3   The test media used, i.e., water, air, or oil, shall be of such a quality that the cleanliness level established at the time of the test shall not be reduced.

7.4   The minimum allowable test medium temperature shall be as follows:

    7.4.1.   Stainless steel and nonferrous material: no restriction.

    7.4.2.   Carbon and low-alloy steel for piping systems (including valves pumps, heat exchangers, etc.) whose associated piping has the following wall thickness:

        a)   under ¾ inch: no restriction.

        b)   ¾ inch and greater: 60°F minimum.

7.4.3 ASME I - Boiler Pressure Test: 70°F.

> **Guideline 164**
>
> {These values should be checked before using them in your particular situation. You have to have the equipment ready and available and it can take a lot of energy to heat up a test medium.}

7.5 Pressure shall not be applied on a system until the temperature of the test media and system under test are approximately equal.

7.6 Special precautions for pneumatic tests - Compressed gas is hazardous when used as a testing medium. It is therefore recommended that special attention be paid to the following:

    7.6.1 Pneumatic tests should be performed when a minimal number of personnel are present.

    7.6.2 Warning signs and barriers should be used to exclude unnecessary personnel from the vicinity of the system under test.

    7.6.3 A preliminary inspection for leakage should always be conducted at no greater than 25 psig. Leaks detected should be repaired prior to achieving test pressure.

    7.6.4 The pneumatic pressure for pneumatic tests shall be gradually increased to less than ½ the test pressure, after which the pressure shall be increased in steps approximately equal to ⅒ the test pressure until the required test pressure is reached.

    7.6.5 When a non-breathable gas is used as the testing medium (e.g., nitrogen), the following precautions should be observed:

        a) Occupied areas should be provided with adequate ventilation and oxygen checks shall be made at regular intervals.

        b) Test personnel should be acquainted with the hazards involved.

c)   All enclosed spaces containing portions of the system under test should be evacuated prior to applying test pressure.

d)   Checks for breathable atmosphere should be performed on all closed spaces prior to entry of test personnel.

e)   A safety monitor should be posted at the entry of a closed space while the test is being conducted. Safety devices such as Scott Air Pak or equal breathing apparatus, rope, portable two-radio, etc., shall be available for the safety monitor to function in a non-breathable atmosphere.

f)   When depressurizing a system following a test, the gas should be conducted to the outside atmosphere via ventilation exhaust ducts or other acceptable means.

7.7   Pneumatic tests should always be performed with a non-flammable, nontoxic gas.

7.8   Lockout/Tagout - AP-2, shall be used for isolation of hazardous energy sources. Where valves are used for pressure test boundary isolation, the valve operator must be locked or chained shut with tags advising "DO NOT OPERATE" and indicating the reason why. The adjacent piping and equipment have to be vented.

Note: Control valves should not be used for a pressure test boundary.

7.9   Overpressure protection shall be provided for the system or equipment being tested. The relief valve(s) shall be set to relieve at 6 to 10 percent or 100 psig above test pressure, whichever is less, and shall be tested just prior to pressurizing the system. The settings shall be affixed directly to the relief valve.

7.10   Heat exchanger(s) included in the test boundary shall be protected during testing by opening drains and vents on the side not under test.

7.11   Hydrostatic test pumps shall be controlled by the Pressure Test Supervisor such that no pump used to pressurize oil-hydraulic systems will be used on water systems or vice versa. Prior to connecting the hydro pump to a system, the pump and associated hoses shall be flushed to ensure the pump discharge will meet the cleanliness requirements of the system or piping.

7.12   Piping designed for vapor or gases shall be adequately supported if tested with a liquid.

7.13   Spring hangers shall be blocked as required for hydrostatic tests.

7.14   All expansion joints shall be restrained so as to prevent damage to bellows during pressurization, if required.

7.15   Relief valves shall be sized in such a way that they shall relieve the full capacity of the test pump at the set point.

7.16   Piping, delineated within a balance of system boundary, open ended to atmosphere, such as relief valve discharge lines or drain lines downstream of the last isolation valve, is not required to be tested.

8.0   Procedure

8.1   Vent high points to ensure that air will be displaced while filling a system for hydrostatic test. Pressure shall be applied slowly and held for a period as specified in Section 10 of this procedure. A pressure gauge shall be installed so that it is visible to the person controlling pressure.

> **Guideline 165**
>
> {Make sure the high points are accessible. If you are using a test media such as oil, someone will have to collect the oil in a receptacle as the oil/air mixture escapes from the vent during the filling process.}

8.2   While pressurizing for the test, pressure shall be held at the respective piping design pressure and the piping equipment shall be inspected for leakage. Unacceptable leakage as defined in Section 10 should be corrected. Leakage from valve packing and mechanical joints, which are required to be leak-tight at design pressure, shall be corrected at this time if possible (i.e., drainage of system is not required if leakage does not exceed capability of pressure source).

CAUTION

Do not tighten valve packing or flanges when the system is above design pressure.

8.3   Increase pressure to test pressure. The pressure requirements, duration, and inspection shall be made in accordance with Section 10.

Note: For pneumatic tests, a water/soap mixture or an acceptable alternate may be used to determine origin(s) of leakage.

Following completion of the inspection, the system pressure shall be slowly reduced to atmospheric. Leakage from valve packing and mechanical joints, which could not be corrected while pressurizing system, should be corrected at this time. Also, weld joints, which did not meet the acceptance criteria, shall be repaired and retested.

9.0   Restoration of Equipment

9.1   If equipment requires draining, vent high points and drain slowly to avoid collapsing of tanks or vessels.

9.2   Upon completion of pressure test on equipment and/or piping requiring "wet lay-up," provide venting to prevent temperature pressurizing and add chemical treatment as required. If freezing climate conditions prevail, the system shall be completely drained and air blown dry as required.

10.0   Acceptance Criteria

10.1   ANSI B31.1 Piping - Piping designed to ANSI B31.1 shall be leak tested as required by Section 2.1.1. Inspection of piping during test shall be made by the contractor's inspection group. All piping that requires a code stamp shall be tested in the presence of an Authorized Code Inspector.

10.1.1   Hydrostatic Tests - The hydrostatic test pressure shall be maintained throughout the duration of the test inspection or for a minimum of 10 minutes. The test will be considered acceptable if no leakage is found by inspecting all areas exposed for test as described in Section 6.1.1. Leakage from temporary gaskets and seals, installed for the purpose of conducting the test and which will be replaced later, may be permitted unless the leakage exceeds the capacity of the pressure source to maintain the required pressure for the duration of the test. Other leaks, such as from permanent seals, valve packing, and gasketed joints, may be permitted but should be noted and corrected at reduced pressures. Valve seat leakage should be evaluated individually based on the applicable design specification and corrected as required. In all cases, permissible leakage shall be directed away from the surface of the component to avoid masking unacceptable leakage from other joints.

10.1.2 Pneumatic Tests - The test will be considered acceptable if no leakage exists as defined in Section 10.1.1.

Guideline 166

{As stated above, you have to replace certain equipment and instruments with blind flanges for the hydrotesting. The gaskets at these blinds are not reusable and have to be replaced when the blind flange is removed. Make sure you have a sufficient quantity of the correct gaskets on hand before you start the testing. When purchasing the equipment, get the vendor to supply additional test gaskets.}

10.2   National Fire Code Piping - The test pressures and acceptance criteria for pressure testing of Fire Protection System Piping are included in NFPA.

10.3   Other Piping - Pressure testing and leak testing of piping designed to codes other than those discussed in this document shall be pressure tested in accordance with the applicable code and/or specific test procedure.

10.4   Records and Reports

10.4.1 Complete documentation for a pressure test shall consist of:

a)   Completed Pressure Test Report Form.

b)   Marked up and signed off system flow diagram and other applicable drawings indicating as-built conditions.

c)   Deficiencies found during the test shall be recorded on the data sheet and retest requirements shall be noted.

d)   Signed off copy of Temporary Change Sheet, signifying completion of all temporary changes.

Note: Temporary change(s) required for a subsequent test shall be transferred to the Temporary Change Sheet of the applicable test.

e) Signed off copy of Test Valve Lineup Sheet.

10.4.2 The completed documentation for a system shall consist of:

a)   The above data included in Section 10.4.1 for each test completed on the system.

b)   A marked up as-built flow diagram(s), which shows a compilation of all pressure tests performed on the system, verifying all required pressure testing is completed.

10.4.3 The above documents shall serve as the official records for pressure tests for a system and are to become part of the test records, which are maintained by the contractor until the system is turned over.

Chapter

12

ELECTRICAL COMMISSIONING PROCEDURES

Electrical Test Procedures

ETP-1 Electrical Test Methods and Procedures

ETP-2 AC Electric Motors

ETP-3 Control Circuitry

ETP-4 Batteries and Chargers

ETP-5 Equipment Grounding

Electrical Test Procedure ETP-1
Electrical Test Methods and Procedures

1.0   Purpose and Scope

1.1   To establish a list of electrical checks and tests performed by electrical startup personnel or a qualified electrical test contractor.

1.2   To define the recommended minimum field test program.

2.0   General Test Method

2.1   The International Electrical Testing Association (NETA) Acceptance Testing Specifications, IEEE, and ANSI shall be used as guidelines for all electrical tests. These three will be referred to in the procedures as "NETA." The contractor is to provide the testing forms to collect the required information to meet the noted standards.

> **Guideline 168**
>
> {Before contract award discuss the issue of providing the required forms, as it may cost the contractor money to find or develop them. Unless you have them readily at hand, it is easier to let the contractor supply them.}

2.2   The pretest checks listed herein are performed by test personnel or verified that they have already been completed by construction or contracted testing company personnel.

3.0   Equipment Tests and Checks

3.1   Transformer - Liquid Filled

3.1.1   Visual and Mechanical Inspection

3.1.1.1   Compare equipment nameplate data with single line diagram and report discrepancies.

3.1.1.2   Inspect for physical damage. Inspect impact recorder prior to unloading transformer, if applicable.

3.1.1.3   Verify removal of any shipping bracing after final placement.

3.1.1.4   Verify proper auxiliary device operation.

3.1.1.5   Check tightness of accessible bolted electrical connections in accordance with manufacturer or NETA specifications.

3.1.1.6   Verify proper liquid level in tank and bushings.

3.1.1.7   Perform other specific inspection and mechanical tests as recommended by the equipment manufacturer.

3.1.1.8   Verify proper equipment grounding

3.1.2   Electrical Tests

3.1.2.1   Perform insulation resistance tests, winding-to-winding, and winding-to-ground, using a megohm meter with test voltage output as shown in NETA specifications. Test duration shall be 10 minutes with resistances tabulated at 1 minute intervals to calculate a polarization index.

3.1.2.2   Perform a turns-ratio test between windings at all tap positions.

3.1.2.3 Perform insulation resistance testing on bushings and lightning arrestor in accordance with NETA specifications.

3.1.2.4 Perform individual excitation current tests on each phase.

3.1.2.5 Perform tests and adjustments on the fan, pump controls, and alarm functions, where applicable.

3.1.2.6 Verify proper core grounding if accessible.

3.1.2.7 Sample insulating liquid in accordance with ASTM. Samples shall be tested for:
- Dielectric breakdown voltage
- Acid neutralization number
- Specific gravity
- Interfacial tension
- Color
- Visual condition
- Water content (required on 25 kV or higher voltages and on all silicone-filled units)
- Dissolved gas - Perform dissolved gas analysis (DGA) in accordance with ANSI/IEEE or ASTM
- Combustible gas - Measure total combustible gas (TCG) content in accordance with ANSI/IEEE or ASTM
- Oxygen content - Perform percent oxygen test

3.2 Transformer - Dry Type larger than 100 kVA single-phase or 300 kVA three-phase

3.2.1 Visual and Mechanical Inspection

3.2.1.1 Compare equipment nameplate data with single line diagram and report discrepancies.

3.2.1.2 Inspect for physical damage, cracked insulators, tightness of connections, defective wiring, and general mechanical and electrical conditions.

3.2.1.3   Verify proper auxiliary device operation.

3.2.1.4   Check tightness of accessible bolted electrical connections in accordance with manufacturer or NETA specifications.

3.2.1.5   Perform other specific inspection and mechanical tests as recommended by the equipment manufacturer.

3.2.1.6   Make a close examination for shipping brackets or fixtures that may not have been removed during installation. Ensure that resilient mounts are free.

3.2.1.7   Verify proper core grounding.

3.2.1.8   Verify proper equipment grounding.

3.2.1.9   Thoroughly clean unit prior to testing.

3.2.2   Electrical Tests

3.2.2.1   Perform insulation resistance tests, winding-to-winding, and winding-to-ground, using a megohm meter with test voltage output as shown in NETA specifications. Test duration shall be 10 minutes with resistances tabulated at 1 minute intervals to calculate a polarization index.

3.2.2.2   Perform power-factor or dissipation-factor tests in accordance with the manufacturer's instructions.

3.2.2.3   Perform a turns-ratio test between windings at all tap positions.

3.2.2.4   Perform winding-resistance tests for each winding at nominal tap setting.

3.2.2.5   Perform individual excitation current tests on each phase.

3.2.2.6   Perform AC overpotential tests on all high- and low-voltage windings-to ground. Use test potentials specified to NETA specifications.

3.2.2.7   Perform tests and adjustments for fans, controls, and alarm functions.

3.2.2.8 Verify that the tap-changer is at specified ratio.

3.2.2.9 Verify proper secondary voltage phase-to-phase and phase-to-neutral after energization and prior to loading.

3.3 Switchgear

3.3.1 Visual and Mechanical inspection

3.3.1.1 Inspect for physical, electrical, and mechanical condition.

3.3.1.2 Compare equipment nameplate information with latest one-line diagram and report discrepancies.

3.3.1.3 Check for proper anchorage, required area clearances, physical damage, and proper alignment.

3.3.1.4 Inspect all doors, panels, and sections for paint, dents, scratches, fit, and missing hardware.

3.3.1.5 Verify that fuse and/or circuit breaker sizes and types correspond to manufacturer's drawings.

3.3.1.6 Verify that current and potential transformer ratios correspond to manufacturer's drawings.

3.3.1.7 Inspect all bus connections for high resistance. Use low-resistance ohmmeter, or check tightness of bolted bus joints by using a calibrated torque wrench. Refer to manufacturer's instructions or NETA specifications for proper torque values.

3.3.1.8 Test all electrical and mechanical interlock systems for proper operation and sequencing.

3.3.1.9 Clean entire switchgear using manufacturer's approved methods and materials.

3.3.1.10   Inspect insulators for evidence of physical damage or contaminated surfaces.

3.3.1.11   Verify proper barrier and shutter installation and operation.

3.3.1.12   Verify proper contact lubricant on moving current-carrying parts and appropriate lubrication on moving and sliding surfaces.

3.3.1.13   Exercise all active components.

3.3.1.14   Inspect all mechanical indicating devices for proper operation.

3.3.2   Electrical Tests:

Perform switchgear electrical tests in accordance with NETA requirements.

3.4   Cable Low-Voltage, 600V Maximum

3.4.1   Visual and Mechanical Inspection

3.4.1.1   Inspect cable for physical damage and proper connection in accordance with single-line diagram.

3.4.1.2   Test cable mechanical connections to manufacture's recommended values using a calibrated torque wrench.

3.4.1.3   Check cable color coding with applicable engineer's specifications and National Electrical Code standards.

3.4.2   Electrical Tests

Perform cable electrical tests in accordance with NETA requirements.

3.5   High Voltage Cable

3.5.1   Visual and Mechanical Inspection
3.5.1.1   Inspect exposed sections for physical damage.
3.5.1.2   Verify cable is supplied and connected in accordance with single-line diagram.
3.5.1.3   Inspect for shield grounding, cable support, and termination.
3.5.1.4   Check for visible cable bends against ICEA or manufacturer's minimum allowable bending radius.
3.5.1.5   Inspect for proper fireproofing in common cable areas.
3.5.1.6   If cables are terminated through window-type CTs, make an inspection to verify that neutrals and ground are properly terminated for proper operation of protective devices.
3.5.1.7   Visually inspect jacket and insulation condition.
3.5.1.8   Inspect for proper phase identification and arrangement.

3.5.2   Electrical Tests:
Perform cable electrical tests in accordance with NETA requirements.

3.6   Switches

3.6.1   Visual and Mechanical Inspection

      3.6.1.1   Compare equipment nameplate information with single-line diagram.

      3.6.1.2   Inspect for physical and mechanical condition.

      3.6.1.3   Check for proper anchorage and required area clearances.

      3.6.1.4   Perform mechanical operation tests.

      3.6.1.5   Verify fuse sizes and types are in accordance with drawings.

      3.6.1.6   Check blade alignment.

      3.5.1.7   Check each fuse holder for adequate mechanical support of each fuse.

      3.6.1.8   Inspect all bus or cable connections for tightness by using calibrated torque wrench. Refer to manufacturer's instructions or NETA Table 10.1 for proper torque levels.

      3.6.1.9   Test all electrical and mechanical interlock systems for proper operation and sequencing.

      3.6.1.10   Clean entire switch using approved methods and materials.

      3.6.1.11   Check proper phase barrier materials and installation.

      3.6.1.12   Lubricate as required.

      3.6.1.13   Exercise all active components.

      3.5.1.14   Inspect all indicating devices for proper operation.

3.6.2   Electrical Tests
Perform all switch electrical tests in accordance with NETA requirements.

3.7   Air Switches - Medium Voltage Metal Enclosed

    3.7.1   Visual and Mechanical Inspection

        3.7.1.1   Inspect for physical and mechanical condition.

        3.7.1.2   Compare equipment nameplate information with single-line diagram.

        3.7.1.3   Check for proper anchorage and required area clearances.

        3.7.1.4   Verify that fuse sizes and types correspond to drawings.

        3.7.1.5   Perform mechanical operator tests in accordance with manufacturer's instructions.

        3.7.1.6   Check blade alignments and arc interrupter operation.

        3.7.1.7   Verify that explosion-limiting devices are in place on all holders having expulsion-type elements.

        3.7.1.8   Check each fuse holder for adequate mechanical support for each fuse.

        3.7.1.9   Inspect all bus connections for tightness of bolted bus joints by using calibrated torque wrench. Refer to manufacturer's instructions or NETA Table 10.1 for proper torque levels.

        3.7.1.10   Test all electrical and mechanical interlock systems for proper operation and sequencing.

        3.7.1.11   Clean entire switch using approved methods and materials.

        3.7.1.12   Verify proper phase barrier materials and installation.

        3.7.1.13   Lubricate as required.

        3.7.1.14   Check switchblade clearances with manufacturer's published data.

        3.7.1.15   Inspect all indicating devices for proper operations.

3.7.2   Electrical Tests
Perform air switch electrical tests in accordance with NETA requirements.

3.8   Air Switches - High and Medium Voltage Open

3.8.1   Visual and Mechanical Inspection
3.8.1.1   Inspect for physical damage and compare nameplate data with plans and specifications.
3.8.1.2   Perform mechanical operator tests in accordance with manufacturer's instructions.
3.8.1.3   Check blade alignment and arc interrupter operation.
3.8.1.4   Check fuse link or element and holders for proper current rating.
3.8.1.5   Check interlocks for correct operation.

3.8.2   Electrical Tests
Perform medium and high voltage switch electrical tests in accordance with NETA requirements.

3.9   Circuit Breakers - Low Voltage Insulated-Case

3.9.1   Visual and Mechanical Inspection
3.9.1.1   Check circuit breaker for proper mounting and compare nameplate data to drawings and specifications.
3.9.1.2   Operate circuit breaker to ensure smooth operation.
3.9.1.3   Inspect case for cracks or other defects.

3.9.2   Electrical Tests
Perform low voltage circuit breaker electrical tests in accordance with NETA requirements.

Electrical Test Procedure ETP-2
AC Electrical Motors

1.0   Purpose

1.1   To determine, by functional testing, that all components operate as designed.

1.2   To verify that the motor is suitable for the service intended and installed properly.

1.3   To provide baseline data for future evaluation of AC motors tested during routine maintenance.

2.0   General

AC motors are generally tested uncoupled, where possible, and then coupled. Coupled testing should be with the system conditions as near to normal or design as possible. When circumstances are such that normal operating conditions cannot be attained, the highest load possible shall be placed on the motor so that meaningful test data may be obtained and evaluated and a reasonable evaluation of the motor's adequacy for the intended service may be made.

3.0   Prerequisites and Initial Conditions

3.1   Prior to performing this test, verify that all circuit breakers, fuses, and overload devices have been properly sized, installed, and tested, as required.

3.2   A functional test of the motor control circuit has been completed.

3.3   For coupled operation, system alignment should be checked to prevent fluid or air flow into undesirable locations.

3.4   Communications shall be established between control and monitoring areas during the performance of this test.

4.0   Precautions and Limitations

4.1   Do not "megger" equipment with solid state components.

4.2   Applicable lockout/tagout procedures shall be followed when performing test.

4.3   Coupling guards shall be installed as soon as practical. All jumpers and lifted leads shall be strictly controlled.

4.5   Do not exceed manufacturer's starting limitations.

5.0   Test Equipment

5.1   Clamp-on ammeter

5.2   Insulation resistance tester

5.3   Multimeter

5.4   Phase-rotation meter

5.5   Stopwatch

5.6   Temperature measuring device. Where automatic measuring equipment is installed for the bearings and/or the windings, it should be used.

5.7   Vibration analyzer

6.0   Uncoupled Motor Test Procedure

6.1   Inspect the foundation and ensure that the motor is properly secured.

6.2   Visually inspect the motor for proper grounding.

6.3   Make sure that it is feasible to uncouple the motor and driven equipment.

6.4   Prior to uncoupled operation, coupling halves shall be tied back to prevent them from contacting any rotating members (if necessary).

6.5   Verify by visual inspection or record check that the motor has been lubricated.

6.6   If possible, manually rotate the shaft and check for free rotation. Note any unusual noise or drag effects.

6.7   Where applicable, verify that the bearing cooling water system is operating properly.

6.8   Where grease is used for lubricant, remove grease plug from one side of each bearing to allow for initial grease expansion.

6.9   Where applicable, visually inspect separate oil cooling systems and verify they are in operation. (Fill if required.)

6.10   Perform and evaluate an insulation resistance test in accordance with project test procedures.

6.11   Measure winding resistance and record results.

6.12   Verify that the motor space heater (when installed) is energized when motor is de-energized.

6.13   Verify proper rotation of motor shaft by "bumping" the motor ON/OFF. Correct rotation shall be accomplished by interchanging two phases of the motor leads.

Note: Rotation may also be checked using a phase rotation meter, but if bumping is used, verify that incorrect rotation will not damage the motor. Before stopping the motor, allow inrush current to decay to running current. Before restarting, make sure the shaft has stopped rotating.

6.14   Start the motor and check for proper operation.

6.15   Where applicable, visually verify that the oil rings are properly distributing oil to the bearings.

6.16   Verify the motor space heater (when installed) is de-energized when the motor is energized.

6.17   Record and evaluate all required electrical data.

6.18   During operational testing, note any unusual noises or physical changes.

6.19   Record and evaluate all bearing temperature readings when stable.

Note: Where winding temperature detecting devices are provided, they should also be monitored and the results recorded.

6.20   Record motor vibration measurements. Record results on a Vibration Record Sheet.

6.21   Stop the motor.

6.22   Restore motor/circuit to pretest conditions.

7.0   Coupled Motor Test Procedure

7.1   Prior to coupled operation, verify that the motor has been tested and is acceptable in accordance with the Uncoupled Motor Test Procedure (see Section 6.0).

7.2   Prior to coupled operation, verify visually or by record check that the mechanical equipment alignment has been completed and both motor and driven equipment have been lubricated, as required.

7.3   Where applicable, verify that the bearing cooling water system is operating properly.

7.4   When provided, visually check separate oil systems for proper operation. (Fill if required.)

7.5   Verify that the system is lined up properly in accordance with applicable flow diagrams.

7.6   Verify proper rotation of motor shaft by "bumping" the motor ON/OFF to determine rotation. Correct if necessary.

CAUTION

If unsure of proper rotation, verify that improper rotation will not damage drive equipment or other parts of the system.

Note: Before stopping the motor, allow inrush current to decay to running current. Before restarting, make sure the shaft has stopped rotating.

7.7   Operate the motor and record all required data.

7.8   While running the motor, take note of any unusual noises or physical changes.

7.9   Where applicable, visually verify that the oil rings are properly distributing oil to the bearings.

7.10   Record and evaluate all bearing temperature readings when stable.

Note: Where winding temperature detecting devices are provided, they should also be monitored and the results recorded in the remarks section.

7.11   Record motor vibration measurements.

7.12   Restore motor/circuit to pretest conditions.

8.0   Acceptance Criteria

8.1   Motors shall correctly respond to control and actuation signals, and the associated indicating lights shall properly reflect circuit status.

8.2   Motors must rotate in the proper direction as required by the driven equipment or as indicated in the manufacturer's instruction manual.

8.3   All recorded data for coupled and uncoupled runs shall be evaluated by the test engineer and be compatible with the nameplate data and the design data for the system.

8.4   Motor insulation resistance shall be acceptable in accordance with NETA guidelines.

8.5   The running current of each phase of the motor shall not vary by more than 6% from the arithmetic average of all the phases. Where variance is greater than 6%, an evaluation should be made by engineering as to the causes.

Note: The 6% limit on phase current variance is imposed to avoid excessive heating effects at full load current. This requirement is not for running currents more than 6% below the full load rating, as the heating effects are different.

9.0 References

9.1 NETA Acceptance Testing Specifications for Electric Power Distribution Equipment and Systems.

9.2 IEEE Standard 112A, Test Procedure for Polyphase Induction Motors and Generators.

9.3 IEEE Standard 114, Test Procedure for Single Phase Induction Motors

9.4 IEEE Standard 43, Testing Insulation Resistance of Rotating Machinery

Electrical Test Procedure ETP-3
Control Circuitry Checkout

1.0   Purpose and Scope

1.1   Determine by functional testing that each electrical control circuit operates, interlocks, controls, indicates, and alarms in accordance with the latest approved elementary, logic, and wiring diagram(s).

1.2   Document the initial energization or retesting of each control circuit and its proper performance.

1.3   This guideline does not apply to instrumentation sensing circuits.

2.0   Definitions

2.1   Positive Logic Verification:
Verification that the presence of all permissives is necessary for the operation of a control function.

2.2   Negative Logic Verification:
Verification that the absence of any one required permissive necessary for positive logic verification will prevent the operation of a control function.

3.0   General

Prior to energizing any control circuit, all of the components in that circuit should be checked for proper installation and adequacy for the intended service. All field wiring should be checked for proper termination continuity and markings. When energized, the circuit should be functionally tested to verify that each control device operates in accordance with the latest approved elementary diagram(s). Elementary diagrams shall be used to perform control circuitry checkout. All logic strings should be checked independently using both positive and negative logic verification. If the logic is very complex, a testing matrix may be used to ensure that all logic strings are tested. Simulation of operation by lifting wires, adding jumpers, or providing separate input devices may be used to verify the circuit.

4.0   Prerequisites and Initial Conditions

4.1   All circuits should initially be de-energized. Both the load equipment and the control circuit shall be electrically isolated from the prime and control power sources. (Remove overloads, place breakers in "TEST" position, or lift loadside cables as applicable.)

4.2   All test equipment and tools shall be within calibration and available at the start of the test.

4.3   All breakers, thermal overloads, and protective relays have been tested and have had their proper settings applied, as required.

4.4   Prior to performing this test, all cables shall have their physical integrity and proper termination verified.

4.5   Prior to performing the test, the Start-up Engineer shall have jurisdictional control over the circuit to be tested.

5.0   Precautions and Limitations

5.1   Applicable tagging procedures shall be followed to ensure personnel safety.

5.2   Communications shall be established as required between control, monitoring, and power source areas.

5.3   Safety equipment, barriers, and warning signs shall be used as necessary to prevent shock hazard from exposed terminals or wires.

5.4   All jumpers and lifted leads shall be strictly controlled. In addition, lifted leads and jumpers that must remain after the test is complete shall be noted in the control room lifted lead/jumper log and on the elementary diagrams.

5.5   All data blanks should be filled in even if not applicable, in which case "N/A" should be used.

5.6   Prior to operating relays or control switches, verify that no other circuit or equipment will be adversely affected by operations.

5.7   Variations from rated control circuit voltage greater than ±10% shall be investigated to determine the cause, e.g., excessive voltage drop, high equalizing charge on DC system, faulty measuring devices, etc. Causes shall be noted.

5.8   If conditions occur during testing which could endanger safety of personnel or damage equipment, the test should be stopped and events reported to the Startup Engineer.

6.0   Test Equipment

6.1   Multimeter

6.2   Timing device (digital timer or stopwatch)

7.0   Procedure

7.1   Visually verify that all control circuit wiring is in accordance with the latest approved elementary and wiring diagram(s).

7.2   Verify that all terminations are right and labeled correctly.

7.3   Verify that each control device and/or circuit component is suitable for use at the voltage level of the circuit (verify nameplate rating).

7.4   Verify that the contact development of all control switches and auxiliary contacts (spares included) is in accordance with the elementary diagram.

7.5   Verify that auxiliary relay contacts (spares included) are in accordance with the latest elementary diagram(s) and terminations are tight.

7.6   Verify the correct type and setting for time delay relays.

7.7   Manually operate auxiliary relays to ensure smooth operation.

7.8   Manually operate all contactors to verify smooth operation and that any mechanical interlocks on the contactor function properly.

7.9   Verify that all fuses, breakers, and thermal overloads are the correct size and type.

7.10   The load should be isolated from its power source (see step 4.1). Energize the control circuit and verify the voltage and polarity of the control power source are as indicated on the schematic drawing.

7.11   Manipulate, either manually or electrically, each control device shown on the elementary diagram to verify that it energizes or de-energizes the component(s) in controlling end device(s). Verify each logic string using both positive and negative logic verification.

7.12   Verify correct indication of all lights, alarms, horns, and computer points.

7.13   During the functional test, "red line" the elementary diagram, note all changes and on completion of test, sign and date it.

7.14   Restore the circuit to pretest condition.

8.0   Acceptance Criteria

8.1   All circuit components shall be suitable for nominal service voltage of the circuit and shall be installed in accordance with the design documents.

8.2   All breakers, thermal overloads, and protective relays have had their proper settings applied.

8.3   Functional testing of the control circuit shall verify that it operates in accordance with the design requirements as delineated in the latest approved elementary, logic, and wiring diagrams.

8.4   Measured control circuit voltages shall be rated nominal voltage ±10% and all variations which exceed these limits must be investigated and the results shown on the data sheet.

8.5   Operating times for time delay relays shall be as shown on the elementary diagrams with allowances for the manufacturer's tolerances for the device.

9.0   References

9.1   Applicable elementary, logic, and wiring diagrams for the circuit under test (including vendor drawings).

9.2   Vendor service bulletins which may be applicable to the equipment under test.

Electrical Test Procedure ETP-4
Batteries and Chargers

1.0   Purpose and Scope

1.1   To verify proper installation of the batteries.

1.2   To verify proper installation and operation of the battery chargers.

1.3   To verify proper installation of the DC panels.

2.0   General

All battery inspections should be made under normal float conditions. Specific gravity readings are not accurate during charge or following the addition of water.

3.0   Precautions and Limitations

3.1   Obtain the proper protective tagging when necessary.

3.2   Ensure that battery room ventilation is operational prior to energizing battery chargers.

3.3   Ensure that portable or stationary safety showers for rinsing eyes and skin are available.

3.4   Ensure that proper protective clothing and face shields are available for worker protection.

3.5   Do not use any conductive material near batteries. Insulate handles of tools used for tightening connector bolts.

3.6   Smoking, arcing, and open flame in the immediate vicinity of the batteries is prohibited.

3.7   Follow the manufacturer's recommendations when energizing or de-energizing battery chargers.

4.0   Inspection of Batteries

4.1   Verify that battery room ventilation is operable.

4.2   Verify that a portable or stationary eye rinse station is in place and operable.

4.3   Verify that intercell connections are correct, clean, and tightened to manufacturer's recommended torque value.

4.4   Inspect the battery for damage, cleanliness, cracks in cell, or leakage of electrolyte.

4.5   Check individual cell electrolyte level; add water as required to bring electrolyte level to the high-level line. Quality is to be in accordance with manufacturer's instructions. (Workers are to wear the proper PPE when filling batteries.)

4.6   Verify cell vent plugs are explosion-resistant types.

4.7   Verify proper installation of the battery rack.

4.8   Verify a small amount of manufacturer's protective grease is applied to connections.

4.9   To remove electrolyte spillage use bicarbonate of soda or other suitable neutralizing agent and a water-moistened wiper.

5.0   Inspection of Charger

5.0   Inspect charger interior and exterior for signs of damage or missing components.

5.1   Clean the charger as necessary. Use a vacuum cleaner to remove dust and dirt.

5.2   Check all control and power connections for tightness.

5.3   Verify the charger is properly grounded

5.4   Verify that nameplate information agrees with design requirements.

5.5   Verify polarity of connections at the charger output matches polarity at battery connections.

5.6   Verify absence of grounds on positive and negative conductors.

5.7   Verify the AC power source is phased, identified, tested, and terminated properly.

5.8   Calibrate all meters associated with the charger. Verify proper sizing of DC shunts associated with ammeters.

5.9   Calibrate and test all alarm devices associated with the charger.

6.0   Inspection of DC Panel

6.1   Examine all power bus compartments for electrical clearances, loose debris, exposed control wiring, etc. Clean up the compartments as necessary.

6.2   Verify proper hardware and torque values on all bus connections.

6.3   Verify that the phasing of the bus throughout the panel is in accordance with project drawings and that no rolls or crosses exist.

6.4   Check any factory applied insulation for cracks, scuffs, tears, dirt, and moisture.

6.5   Verify proper grounding of the panel.

6.6   Inspect the panel for proper arrangement of disconnects, alarm devices, and meters. Verify equipment is not missing or damaged.

6.7   Check the operation of mechanical hardware with respect to alignment, hinges, locks, and hold-open devices. Realignment and/or lubrication may be required to provide easy operation.

6.8   Verify proper operation and rating of disconnect/transfer switches and breakers associated with the panel.

6.9   Calibrate all panel meters. Verify proper sizing of DC shunts associated with ammeters.

6.10   Calibrate and test all alarm devices associated with the panel.

6.11   Verify panel is anchored securely.

6.12   After inspection reinstall all power bus coverplates and megger the power bus, disconnect/transfer switches and breakers associated with the bus.

6.13   Verify that the DC power source is phased, identified, tested, and terminated properly.

7.0   Initial Charge and Tests

7.1   Prior to energizing battery, verify correct sequence of AC power source at the charger.

7.2   Energize the battery charger. Adjust the float and equalize voltages and apply initial charge to the batteries per manufacturer's instructions.

7.3   After the battery has returned to float charge for 72 hours measure the specific gravity, temperature, and voltage of each cell.

7.4   Check operation of cooling fans.

7.5   Monitor battery cell electrolyte level. Remove electrolyte as required.

7.6   Correct the specific gravity of each cell electrolyte. For temperature correction add .001 to measured value for each 3°F over 77°F and subtract .001 for each 3°F below 77°F.

7.7   If cell temperatures deviate more than 5°F from each other during inspection, determine the cause and note in remarks sections.

Electrical Test Procedure ETP-5
Equipment Grounding

1.0   Purpose and Scope

1.1   To verify the grounding of electrical equipment according to specified requirements.

2.0   General Test Method

2.1   All electrical equipment will be either inspected visually or by record to verify proper grounding.

3.0   References

3.1   National Electric Code 1994, Article 250.

3.2   Engineer's specified design criteria.

4.0   Special Test Equipment

4.1   Ground megger tester

5.0   Prerequisites

5.1   The associated system file shall be available and the assigned startup personnel shall be familiar with that portion of its contents relative to the grounding system in question.

5.2   Startup engineers must have made a preliminary inspection of the system of equipment and accepted it from construction.

6.0   Precautions and Limitations

6.1   Before any item of equipment is energized, verification must be made that the grounding requirements are complete (i.e., case, structure, etc.).

7.0 Procedure

7.1 The facility and switchyard ground grids may be installed before startup personnel are active at the site. They will be buried in the earth and thus inaccessible for inspection. Their integrity will be based upon construction inspection records and test results.

7.2 In the Plant, verify that structures and equipment are grounded in accordance with design requirements. The NEC grounding requirements are summarized below.

    7.2.1 All Plant electrical equipment shall be grounded. The main process building and all auxiliary building structural steel shall also be grounded.

    7.2.2 Power transformers shall be connected to the ground grid in at least two places with 350 MCM copper cable. Where surge arrestors are provided, the surge arrestor ground connection must be made either to the transformer or directly to the ground grid with electrical conductors provided specifically for ground purposes (not support bolts). If a discharge counter is supplied, manufacturer's instructions should be followed. The transformer neutral requires one additional 350 MCM tap to the ground grid usually from a ground bus to which the neutral bushing is connected.

    7.2.3 In general, the conduit containing the control conductors constitutes the primary ground conductor. When conduit is not used, a ground conductor shall be used.

7.2.4   Any cable tray system shall be continuously grounded throughout the Plant. All sections of cable tray shall be electrically bonded to adjacent sections. Conduit terminating at cable tray shall be bonded to the tray. Ground conductors run in the supply conduit shall be connected to the cable tray flange adjacent to where the conduit is terminated. A ground cable (min. No. 4/0 AWG) shall be laid in all trays containing power cables.

7.2.5   All 575 V motors will use mounting bolts for grounding to structural steel. All 4 kV motors will be grounded using a ground conductor fastened to the motor frame. Motors 25 hp and smaller will use supply conduit or a ground conductor.

7.2.6   All switchgear, motor control centers, and load centers will be grounded by ground cables at each end of the ground bus. Individual loads will be grounded by conduit or ground conductors.

7.2.7   Compression lugs will be used at equipment ground point where practical. Compression lugs will be of 2 hole, 2 dent type where space permits.

7.2.8   Where ground conductors are run to building steel, they will be CAD welded to the building steel.

7.2.9   Instrument cable shields (drain wire) will be grounded at one point only—at the equipment or device (field) junction box. Ties must not touch ground.

7.2.10   Minimum cable size used for miscellaneous grounds and ground jumpers is No. 8 AWG copper.

7.2.11   Verify "quiet" grounding of dedicated computer ground and cable shield grounds.

7.3   Grounding Grid Test

    7.3.1   A grounding grid test megger shall be used to test and verify the grounding grid acceptance.

8.0   Restoration of Equipment

8.1   Restore all equipment to its pretest condition.

9.0   Acceptance Criteria

9.1   Grounding shall be in accordance with Section 7.0.

9.2   Ground conductors shall be equal or greater in ampacity to the power leads.

INSTRUMENTATION COMMISSIONING PROCEDURES

Instrument Test Procedures

ITP-l   Instrumentation and Controls Calibration Methods

ITP-2   Instrumentation and Controls Installation Checkout

ITP-3   Instrumentation and Controls Calibration and Functional Checkout

Instrument Test Procedure ITP-1
Instrument and Controls Calibration Methods

1.0   Purpose

1.1   Establish the method of the instrument and controls calibration.

1.2   Establish procedures for performance of instrumentation and controls calibration and test procedures.

2.0   Scope

2.1   All instrument and controls calibration and checkout will be performed by the startup group using certified calibration instruments and applicable vendor instrument test procedures and Instrument Test Procedure ITP-2 Installation Checkout and ITP-3 Calibration and Functional Checkout Procedures.

2.2   All instrument and controls calibration and checkout will be performed by qualified technicians.

3.0   Responsibilities

3.1   Engineers:
The startup manager is responsible for supervising the I&C calibration and test program. He shall supervise the calibration personnel to ensure that all instruments and instrument loops are checked out in a timely and professional manner and that they are properly documented. He shall coordinate the I&C program activities with the requirements of the overall startup program.

3.2   Calibration Personnel:
The technicians will conduct device calibration and loop checkout activities as directed by the SM.

4.0   Initial Conditions

4.1   All instruments and controls to be tested shall be installed, complete and ready for operation.

4.2   All instruments shall be turned over by the Contractor's construction group to the calibration technicians before the start of instrument calibration and loop checking.

4.3   All test equipment used for calibration shall be calibrated by a certified testing laboratory.

5.0   Precautions and Limitations

5.1   Follow all necessary lockout/tagout procedures when performing calibration of installed devices.

5.2   When operating equipment, good operating practice and safety guidelines will be followed.

6.0   Procedure

6.1   All calibration and checkout will be performed using the applicable vendor test and calibration procedures.

6.2   Local indicators will be bench calibrated and then installed. All other instruments and instrument loops will be calibrated after installation is complete.

6.3   Installation checkout will be performed using ITP-2, Instrumentation and Controls Installation Checkout Procedure.

6.4   Instrument calibration and functional checkout will be performed using ITP-3, Instrumentation and Controls Calibration Functional Checkout Procedure.

7.0   Acceptance Criteria

7.1   Acceptance criteria shall be as outlined in Section 9.0 of ITP-3.

Instrument Test Procedure ITP-2
Instrumentation and Controls Installation Checkout

1.0   Purpose

1.1   To verify that instruments and associated piping, mounts, racks, and accessories are installed according to the latest approved installation drawings and that the installation conforms to good construction practice.

1.2   To verify that instrumentation electric and electronic circuits are wired correctly.

2.0   General Test Method

2.1   A visual inspection of all instrumentation and associated piping and wiring as shown on the installation details and cable schedules shall be performed and documented.

3.0   References

3.1   Approved vendor drawings.

4.0   Special Test Equipment

4.1   Multimeter

4.2   Continuity tester

5.0   Prerequisites

5.1   The instrumentation loop has been determined complete by the construction group and turned over to the startup group.

5.2   All required tools, instruments, and other devices shall be available prior to the start of a specific instrument loop check.

6.0   Precautions and Limitations

6.1   Follow all lockout/tagout procedures while conducting tests.

6.2   Use suitable voltage-detecting devices to verify that the equipment to be energized is not energized.

7.0   Procedure

7.1   Verify the proper instruments are installed and auxiliary equipment, e.g., solenoid valves, position switches, positioners, air sets, etc., are properly installed, flow elements properly oriented, and process pressure taps piped to the correct instrument ports, in accordance with the installation details.

7.2   Verify all instruments are installed to allow access for calibration and maintenance, all required test connections are properly installed, and instruments and tubing are adequately supported.

7.3   Verify all loop components have been properly and permanently installed, all flex and conduit connections are complete and fittings are secure, all panel and device covers are properly installed, drain lines are complete, all equipment and cables are properly identified, and all components are free from visual damage and are clean of objectionable materials.

7.4   Verify the instrument primary sensing and pneumatic piping has been checked for continuity, blowndown, and leak tested, or perform these tests at this time. Refer to Instrumentation Installation Specifications for blowdown procedures. Verify the instrument sensing lines slope in the proper direction, in accordance with applicable standards.

7.5   Verify all wiring terminations are correct, as indicated on the cable schedule and vendor wiring diagrams.

7.6   Verify all instrument devices are identified with instrument mark number stamped on metal tags.

8.0   Restoration

8.1   All equipment shall be restored to its pretest condition.

9.0   Acceptance Criteria

9.1   All loop devices shall conform to applicable drawings and specifications for intended function, correctness of mounting, supporting, routing, tubing slope, device location, and good workmanship practices.

Instrument Test Procedure ITP-3
Instrumentation and Controls Calibration and Functional Checkout

1.0   Purpose and Scope

1.1   To verify all instruments are calibrated in accordance with approved procedures.

1.2   To determine, by functional testing, that an instrument loop is complete and that satisfactory response of the equipment is obtained.

1.3   To provide baseline data for the evaluation of future testing of the instrumentation.

2.0   General Test Method

2.1   Individual instruments will be calibrated by an authorized calibration contractor.

2.2   Instrument loops will be checked functionally to verify proper response of equipment.

3.0   References

3.1   Applicable instrument technical data sheet

3.2   Applicable manufacturers' instruction manuals

3.3   ITP-2, Instrumentation and Controls Installation Checkout Procedure

4.0   Special Test Equipment

4.1   Test and calibration equipment as required by applicable instrument calibration procedures and manufacturer's instructions.

4.2   All test equipment and devices shall be calibrated prior to starting the calibration program.

5.0   Prerequisites

5.1   ITP-2, Instrumentation and Controls Installation Checkout, shall have been satisfactorily completed, where applicable.

5.2   All required tools, instruments, and other devices, properly calibrated, shall be available prior to the start of a specific instrument loop checkout.

6.0   Precautions and Limitations

6.1   AP-2, Lockout/Tagout Procedure, shall be followed while conducting tests.

6.2   Assure that operation of equipment during testing will not create plant disturbances or personnel hazards.

7.0   Procedures

7.1   Device Calibration

   7.1.1   Using the vendor's instrument calibration procedures and/or manufacturer's instructions, calibrate all instruments.
   Note: On pressure instruments, make head corrections as necessary and record on the Instrument Calibration Data Sheet.

   7.1.2   All instrument calibration set-points and ranges shall be obtained from the engineer's Instrument Calibration Data Sheet when available or from the manufacturer's instruction manual.

7.1.3   All instrument device calibration data (set-points/ranges) shall be recorded on the appropriate Instrument Calibration Data Sheet.

7.1.4   After each instrument is calibrated, a calibration "sticker" shall be placed on the instrument for verification of calibration.

7.2   Functional Loop Checkout

7.2.1   Energize the circuit(s) and turn on all air supplies as required for functional testing.

7.2.2   Analog and Digital Loops
Simulate the process variable to actuate the primary device and observe the loop response. Record on the loop calibration sheet the indication on all indicating and recording devices, valve/damper positions, contact closures, and any other information required.

For digital loops, computer and alarm circuits shall be verified and recorded.

7.3   Record the test equipment used and all observed data on the Loop Calibration Data Sheet.

8.0   Restoration of Equipment

8.1   The instrument loop shall be restored to the normal operating condition.

9.0   Acceptance Criteria

9.1   All signal levels, both electrical and pneumatic, shall be in agreement with the vendor design drawing and instrument data sheets.

9.2   The calibration accuracy of all instrumentation shall be in agreement with the instrumentation specification and data sheets.

9.3   Signing the appropriate block in the Loop Calibration Data Sheet shall indicate satisfactory completion of Paragraph 7.2.

Conclusion

The system is now commissioned and is ready for startup. It is now turned over to operations for the introduction of raw material and full production. If there are performance guarantees, there will now be a lead-up period to the start of them. Your roll at this point is that of an observer. If there are operations problems, you should investigate them thoroughly. You will have to determine if it is an equipment/system design fault or an operator training issue.

The last part of your project is the close-out. This is making sure that all FWOs are closed, Substantial and Final Completion forms have been issued, and all your paperwork is complete. You can strip your files and keep the information you feel is relevant for future use. Check with your purchasing group to see how long the information should be kept on hand—usually seven years. Most plants have a dead file area where your material will be stored.

Now that you have finished this book you have a good basis for further studies in project management. I hope you have acquired some useful information and are able to put it into practice. If this book has helped you understand in more detail what you should be doing and why, then the book has met its objective.

Subject Index

A

B

D

E

F

G

H

I

K

L

M

Mechanical Commissioning
    see Commissioning, Procedures, Mechanical
Mechanical Completion, 281, 293
Metric, 79-81, 104, 110
Minutes of Meeting, 39, 240
Mobile Equipment, 40

N

North Arrow, 79-80,
Non-Destructive Testing (NDT)
    see Estimates, Piping, Testing and Inspection
Notebook, 3, 310

P

Permits, 29, 51, 54, 62, 168, 204, 219
    Building, 30, 54, 168
    Environmental, 30, 50, 54
Photographs, 167, 269
Pipe Insulation, 44
Pipe Penetrations, 44
Piping, 41, 111, 188, 333, 341
    Construction Completion Phase, 298
    Pressure, 42
Piping Requirements, 41
Piping Tie-Ins, 42
Plant Hazards, 65
Plant Capital Budget
    see Budget, Capital
Plant Maintenance Budget
    see Budget, Maintenance
Plant Overhead Budget
    see Budget, Overhead

S

T

U

V

W

ABOUT THE AUTHOR

Morley H Selver earned his Bachelor of Science degree in Civil Engineering in 1973 from the University of Manitoba. He has over 25 years of experience in industrial plant project engineering. He has worked in industry, with and for consultants, and had his own consulting business. This project experience includes operations and maintenance, research and development, project management of small to medium size plant projects, construction management of large industrial projects, mechanical installation of heavy industrial equipment, commissioning and startup of industrial plants, and plant management.

He has worked in operating pulp and paper mills, oil refineries, on North Slope oil projects, board plants, and in the recycling industry. These plants were located in various locations across Canada and the USA. He is currently working in the consulting business.

Did you like this book?

If you enjoyed this book, you will find more interesting books at

www.MMPubs.com

Please take the time to let us know how you liked this book. Even short reviews of 2-3 sentences can be helpful and may be used in our marketing materials. If you take the time to post a review for this book on Amazon.com, let us know when the review is posted and you will receive a free audiobook or ebook from our catalog. Simply email the link to the review once it is live on Amazon.com, with your name, and your mailing address—send the email to orders@mmpubs.com with the subject line "Book Review Posted on Amazon."

If you have questions about this book, our customer loyalty program, or our review rewards program, please contact us at info@mmpubs.com.

"Simply put, this book is about helping people control smaller projects in a logical and effective way, and making the process run smoothly, and it is indeed a success in achieving that goal."

IEE Engineering Management (12/04)

Mike Watson

Award-winning trainer and PM Methodology guru

Managing Smaller Projects: A Practical Approach

So called "small projects" can have potentially alarming consequences if they go wrong, but their control is often left to chance. The solution is to adapt tried and tested project management techniques.

This book provides a low overhead, highly practical way of looking after small projects. It covers all the essential skills: from project start-up, to managing risk, quality and change, through to controlling the project with a simple control system. It cuts through the jargon of project management and provides a framework that is as useful to those lacking formal training, as it is to those who are skilled project managers and want to control smaller projects without the burden of bureaucracy.

Read this best-selling book from the U.K., now making its North American debut. *IEE Engineering Management* praises the book, noting that "Simply put, this book is about helping people control smaller projects in a logical and effective way, and making the process run smoothly, and is indeed a success in achieving that goal."

ISBN: 9781895186857   (paperback)

Also available in ebook formats. Order from your local bookseller, Amazon.com, or directly from the publisher at
http://www.mmpubs.com/

Lessons from the Ranch for Today's Business Manager

The lure of the open plain, boots, chaps and cowboy hats makes us think of a different and better way of life. The cowboy code of honor is an image that is alive and well in our hearts and minds, and its wisdom is timeless.

Using ranch based stories, author Michael Gooch, a ranch owner, tells us how to apply cowboy wisdom to our everyday management challenges. Serving up straight forward, practical advice, the book deals with issues of dealing with conflict, strategic thinking, ethics, having fun at work, hiring and firing, building strong teams, and knowing when to run from trouble.

A unique (and fun!) approach to management training, Wingtips with Spurs is a must read whether you are new to management or a grizzled veteran.

ISBN: 1-897326-88-2   (paperback)

Also available in ebook formats. Order from your local bookseller, Amazon.com, or directly from the publisher at
http://www.mmpubs.com/

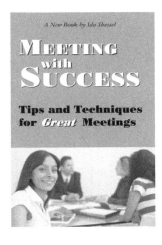

CPSIA information can be obtained at www.ICGtesting.com
Printed in the USA
LVOW01*2034300415

436820LV00005B/46/P